WEIGHTLESSNESS—
PHYSICAL PHENOMENA
AND BIOLOGICAL EFFECTS

Norair Zero Gravity Tower

AN AMERICAN ASTRONAUTICAL PUBLICATION
SOCIETY

WEIGHTLESSNESS— PHYSICAL PHENOMENA AND BIOLOGICAL EFFECTS

Edited by

ELLIOT T. BENEDIKT

Proceedings of the Symposium on
Physical and Biological Phenomena
Under Zero G Conditions
Held July 1, 1960 at Hawthorne, California

Springer Science+Business Media, LLC

ISBN 978-1-4899-6152-5   ISBN 978-1-4899-6279-9 (eBook)
DOI 10.1007/978-1-4899-6279-9

© Springer Science+Business Media New York 1961

Originally published by American Astronautical Society, Inc. in 1961.

Softcover reprint of the hardcover 1st edition 1961

First Printing

Library of Congress catalog card number: 61-15157

FOREWORD

The symposium, "Physical and Biological Phenomena Under Zero G Conditions," was sponsored by the American Astronautical Society in conjunction with the Norair Division of the Northrop Corporation. The second joint AAS/industry professional meeting in the astronautical sciences, this symposium was designed to present theoretical and experimental results of zero g research, to discuss the design of instrumentation required for further exploration in this field, and to encourage scientists and engineers to give increased attention to research and development aspects of this unique space environment.

As a member of the Board of Directors of the American Astronautical Society, I want to express appreciation to the Norair Division of the Northrop Corporation and to Dr. Elliot T. Benedikt, Technical Sessions Chairman, for their significant contributions to this symposium.

NORMAN V. PETERSEN, PROGRAM CHAIRMAN
Chief, Astro Sciences Group
Norair Division, Northrop Corporation

PREFACE

In organizing the symposium, "Physical and Biological Phenomena under Zero G Conditions," an attempt was made to assemble a representative cross-section illustrating the present status of scientific and technological knowledge in this new and rapidly growing field. The very newness of the field posed a number of difficulties. First, the task of obtaining a fairly complete representation of all significant contributions was made difficult by the lack of a body of published literature. Second, results in a new branch of knowledge are bound to be tentative and sometimes in need of revision in the light of subsequent developments. These difficulties could have been avoided by waiting for additional, better publicized and critically assessed results. However, an early interchange of ideas and the stimulation resulting from the dissemination and exchange of information, even though preliminary, appeared to be more beneficial than a later collection of possibly more advanced and better settled views. This series of papers is offered to the interested scientific public in this spirit.

In assembling the proceedings of the symposium, the Editor has endeavored to provide a systematic sequence of topics, with the intent that this volume fulfill at least partially the functions of a handbook on the subject, until such becomes available. Thus, the order of the various contributions is different from their actual scheduling in the symposium.

The Editor wishes finally to mention that he was greatly assisted in the preparation of this publication by his wife, Geraldine P. Benedikt, and his associates, Messrs. R.W. Halliburton and R. Lepper.

ELLIOT T. BENEDIKT, EDITOR
Head, Space Physics Laboratory
Astro Sciences Group
Norair Division, Northrop Corporation

CONTENTS

PART ONE

Hydrodynamics at Zero Gravity

GENERAL BEHAVIOR OF A LIQUID IN A ZERO OR NEAR-ZERO GRAVITY ENVIRONMENT

ELLIOT T. BENEDIKT

Head, Space Physics Laboratory, Astro Sciences Group,
Norair Division, Northrop Corporation, Hawthorne, California

In the absence or near-absence of external forces—in particular, gravity or equivalent inertial forces—the behavior of liquid masses partially or totally bounded by free surfaces is governed exclusively by intermolecular actions, such as cohesion, adhesion and gravitational attraction. Available knowledge regarding the physics of liquids under such conditions appears to be limited largely to the statics of thin liquid films or systems of small dimensions such as bubbles or drops.

Situations in which liquids bounded (at least partially) by free surfaces and upon which gravitational or inertial forces cease to act, or do not act at all, are frequently encountered in space technology and space medicine. As examples, we might cite the problem of fuel storage and supply in missiles and space vehicles, the behavior of liquid metals in welding operations in space and, of course, the important physiological effects due to the lack of gravity.

Apart from these and other applications, the subject seems to offer many aspects of fundamental interest for the physics of liquids. Some of the mathematical, physical and experimental problems connected with this question will be discussed in what follows.

Intermolecular Forces

It will be appropriate to commence our discussion with a review of molecular interactions which, as mentioned above, determine the behavior of a liquid in a weightless or near-weightless condition.

Interactions between molecules can be described in terms of short-range valence forces and longer-range so-called Van der Waals forces. The valence forces are called into play when the electronic

3

structures of the molecules overlap and are responsible for the apparent impenetrability of matter.   Their intensity decreases very rapidly—approximately exponentially—with the intermolecular distance.   The small Van der Waals forces are due in part to the conventional electrostatic repulsion and attraction between the individual electrons and nuclei and in part to quantum mechanical effects.   It is customary to subdivide these forces according to their origin, i.e., into (I) electrostatic forces, due to the interaction between permanent electric dipole moments; (II) induction forces, due to the interaction between permanent and induced electric multipole moments; (III) dispersion forces, arising from the interaction between instantaneous electric multipole moments; and (IV) resonance forces, originating as a consequence of photon exchanges between molecules.   Whereas the laws of the above interactions are considerably complicated, it is possible to describe their statistically averaged effects in terms of an attraction derivable from a potential of the type $C/r^6 + \ldots$, r being the intermolecular distance, and the dots representing additional terms inversely proportional to higher powers of r.   For water, the coefficient C assumes the following expressions and numerical values:

$$C = (2/3)\, \mu^4/kT = 19.0 \times 10^{-59}\, (T_0/T)\ \text{erg cm}^6$$

$$C = (3/4)\, a^2\, h\nu_0 = 4.7 \times 10^{-59}$$

$$C = (3/4)\, a\mu^2 = 1.0 \times 10^{-59}$$

respectively for electrostatic, dispersion and induction interaction forces.   In the above formulae, $\mu = 1.84 \times 10^{-18}$ e.s.u. represents the permanent dipole moment of a water molecule, $a = 1.48 \times 10^{-24}$ cm^3 its polarizability, $h\nu_0 = 2.88 \times 10^{-11}$ erg its molecular vibration energy. T is the absolute temperature and $T_0 = 293°$K.

A better idea of the magnitude of intermolecular forces can be obtained from Table 1, in which estimates of the over-all forces exerted between two water molecules at various intermolecular distances are presented (the radius r_{H_2O} of the water molecule having been identified with the cube root, $\sqrt[3]{a} = 1.144 \times 10^{-8}$ cm of its polarizability).   For purposes of comparison, it can be noted that a molecule of water weighs 3×10^{-20} dyne on the surface of the Earth.

Surface Tension

Cohesion and adhesion. Let us consider now two differential elements of volume dV_1, dV_2 (see Figure 1a) of a liquid separated by a distance r_{12}. If N is the particle density (supposed to be uniform) of the liquid, these two elements will contain NdV_1, NdV_2 molecules respectively. Hence their interaction energy will be $N^2\Phi(r_{12})dV_1 dV_2$ where Φ represents the statistically averaged interaction energy between the two molecules. The total interaction energy will therefore be

$$U = (1/2) \int_V \int_V N^2 \, \Phi \, (r_{12}) \, dV_1 dV_2$$

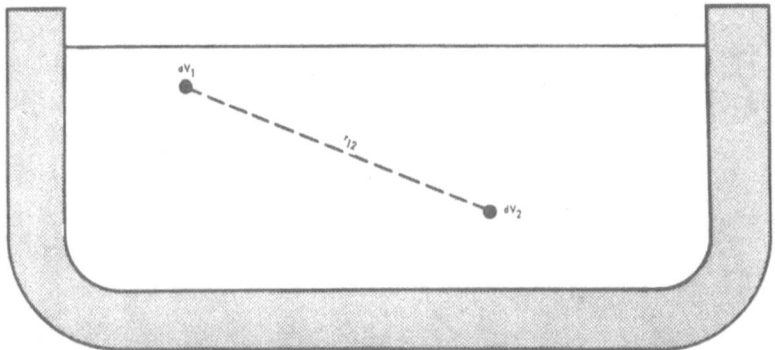

Figure 1a. Interaction between molecules of liquids as cause of cohesion.

By standard procedures of mathematical physics (the details of which can be found in Ref. 1), the two successive volume integrations can be expressed as a combination of double volume and surface integrals. Utilizing the shortness of the range of intermolecular forces (discussed in the preceding article) and introducing the two auxiliary functions

$$\psi(r) = \int_r^\infty r'^2 \, \Phi(r') \, dr', \qquad \chi(r) = \int_r^\infty \psi(r') \, dr'$$

the result can be expressed in the form

$$U = -4\pi N^2 \, \psi(0) \, dV + \pi N^2 \, \chi(0) \, A + \ldots$$

The term omitted becomes important only when the liquid system, or a part of it, has the shape of a very thin membrane; otherwise it is negligible. The two terms in the above formula represent thus the contribution of the intermolecular or cohesional energy to the internal energy of the substance. Similarly, in the case of interaction between two media, a and b, in contact (as in the situation depicted in Figure 1b, of a liquid bounded by a container) the potential energy of the attraction between the molecules of the two media, i.e., the energy of cohesion, will be given by

$$U^{(a,b)} = \int \int N^{(a)} N^{(b)} \; \phi^{(a,b)} \; (r_{12}) \; dV_1{}^{(a)} dV_2{}^{(b)}$$

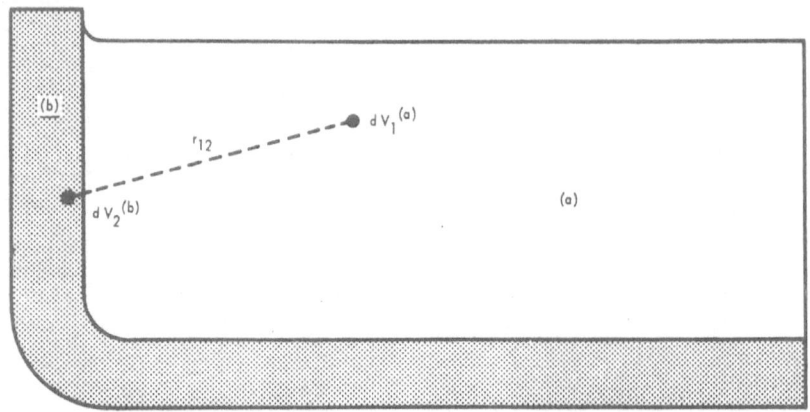

Figure 1b. Interaction between molecules of solid container and liquid as cause of adhesion.

In the above formula, $N^{(a)}$, $N^{(b)}$ represents the density of the particles in the media, $dV_1^{(a)}$, $dV_2^{(b)}$ infinitesimal elements of volume in these media surrounding two points at an intermolecular distance r_{12} and $\phi^{(a,b)}$ the average potential energy of particles of the two media separated by the above distance. Introducing the function

$$\chi^{(a,b)}(r) = \int\limits_{r}^{\infty} \int\limits_{r'}^{\infty} r'' \; \phi^{(a,b)} \; (r'') \; dr'' \, dr'$$

we can write (see Ref. 1, loc. cit.)

$$U^{(a,b)} = -\pi \, N^{(a)} N^{(b)} \; \chi^{(a,b)} \; (0) \; A^{(a,b)} + \dots$$

TABLE 1: FORCES BETWEEN WATER MOLECULES

Intermolecular distance

	$1r_{H_2O}$	$10r_{H_2O}$	1μ
Valence repulsion	4.7×10^{-2} dyne	1.31^{-21} dyne	—
Van der Waals attraction	4.7×10^{-2}	4.7×10^{-9}	1.2×10^{-29} dyne

where terms which are negligible, unless some portions of the media are extremely thin, have been omitted.

It appears thus that the potential energy of the attraction between the molecules of two media, i.e., the energy of cohesion is proportional to the area $A^{(a,b)}$ of the surfaces in contact.

A summary of the more recent developments regarding an explanation of surface tension from the point of view of molecular theory can be found in Ref. 2.

Thermodynamic definition of surface tension. It is apparent from the above developments that a portion of the cohesive energy can be regarded as distributed over the surface of bodies, in particular of liquids, with a certain energy density u_*. However, since most experimentation is conducted at constant temperature, it is preferable to introduce the concept of surface tension σ, defined as free surface energy. The operational significance of this definition can be illustrated by use of a device (illustrated in Figure 2)

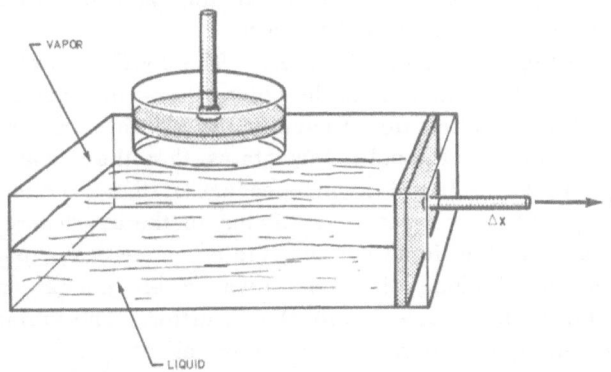

Figure 2. Idealized method for production of isochoric and isothermal transformation of liquid-vapor system, illustrating thermodynamic definition of coefficient of surface tension as surface density of free energy.

consisting of a closed container with one movable wall, partially
filled with a liquid and surmounted by a cylinder terminated by a
weightless piston. The system will be regarded as maintained at
a constant (absolute) temperature T, and subjected to a uniform
gravitational field of intensity g; moreover, it will be assumed that
the forces exerted upon a molecule of the liquid by the walls of the
container are equal to those exerted upon it by the remaining molecules of the liquid. By means of this idealized arrangement, it is
possible to vary, by appropriate simultaneous displacements of the
pistons, the area of the surface of the liquid without changing the
volume V of the system—that is, without work being contributed by
pressure forces. Let us indicate by F the force exerted (normally)
upon the movable vertical wall of the container, Δx its displacement, and with $\Delta \mathbb{W} = F \Delta x$ the work thus performed. The coefficient
of surface tension can then be defined as

$$\sigma = \lim_{\Delta x \to 0} \left[\left(\frac{\Delta \mathbb{W}}{\Delta A} \right)_{V,T} - m g \frac{\Delta \bar{z}}{\Delta A} \right]$$

i.e., the nonmechanical portion of the work required under such
conditions to increase the area of the liquid by unity. In the above
expression, m denotes the mass of the liquid, $\Delta \bar{z}$ the change in the
height of its center of mass.

The coefficient of surface tension thus essentially measures the
work which must be performed in bringing (isothermally) molecules
from the interior to the surface of the liquid, when the area of the
surface of the latter is increased by unity.

Kinetic definition of surface tension. The coefficient of surface
tension can also be defined from the point of view of kinetic theory,
in terms of the deviation from Pascal's law which occurs near the
surface of a liquid. Pascal's law states that the pressure at a
given point inside a liquid is independent of the direction along
which this pressure is measured. Owing to the asymmetry of cohesive forces near the liquid surface, the components p_N, p_T of the
pressure measured along directions respectively normal and tangential to the surface (see Figure 3) will differ. The surface tension can accordingly be defined by the integral

$$\acute{\sigma} = \int_{z_1}^{z_2} \left[p_N - p_T(z) \right] \, dz$$

which must be regarded as extended between the ordinates of points placed within a distance equal to a few molecular radii within the liquid and its vapor respectively. As already implied, the transition layer, in which the field of force acting on the molecules of the liquids can be expected to be different from the fields prevailing in the interior of the substance, appears to extend over a thickness of a few molecular radii only. This was experimentally demonstrated by the analysis of the reflection of a polarized beam of light from the liquid surface.[3] Hence the above kinetic definition, which involves the definition of pressure forces acting over areas of molecular dimensions, is somewhat questionable.

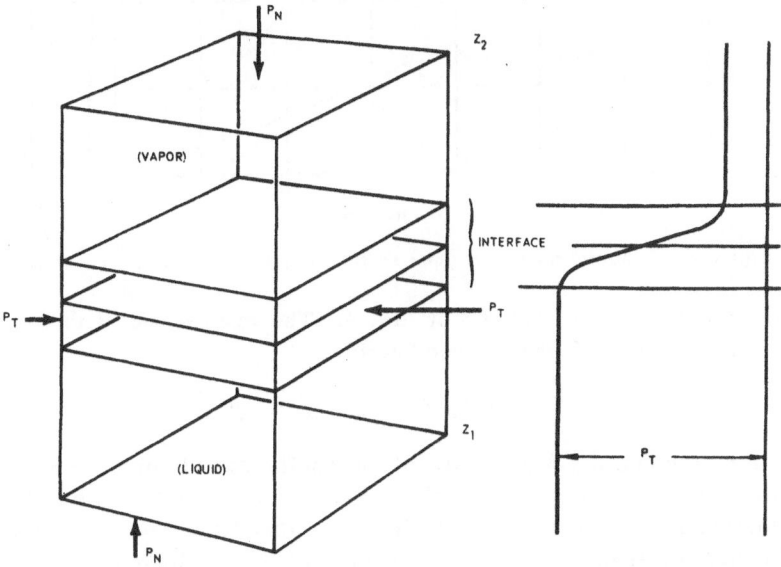

Figure 3. Transition of tangential pressure through liquid-vapor interface, illustrating kinetic origin of surface tension.

Numerical values of the coefficient of surface tension. Numerical values of the coefficient of surface tension for various typical liquids are shown in the first column of Table 2. The coefficient of surface tension—i.e., the surface density of free energy—and the surface tension energy, or the surface densities of the contribution of the surface to the internal energy of the liquid, are represented for water in Figure 4, which was constructed on the basis of data

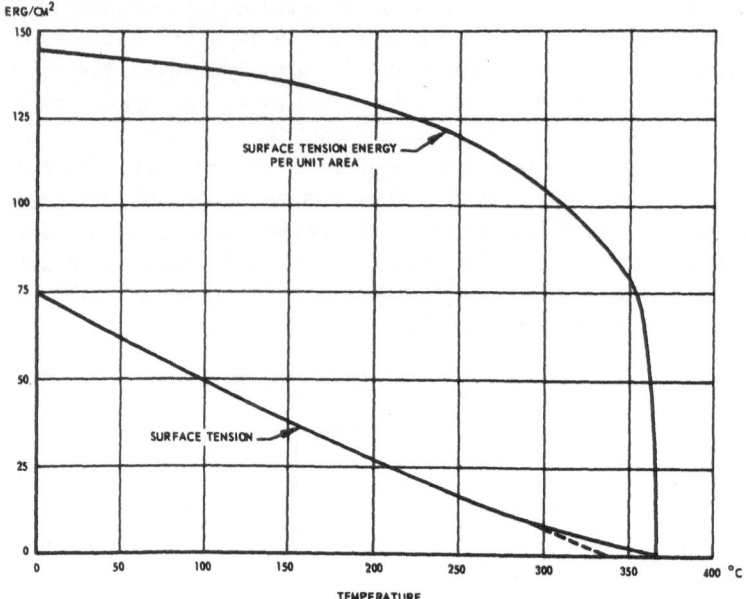

Figure 4. Temperature effect upon surface tension and energy for water.

appearing on pages 262-263 of Ref. 4. The results shown there il-
lustrate the law of Ramsay and Shields,

$$\sigma = (K/\omega^{2/3}) \ (T - T_* - d)$$

which provides the dependence of the coefficient of surface tension
upon the temperature. In the above formula, ω represents the molecular volume of the liquid and T_* its critical temperature at which
the surface tension and surface energy of the liquid reduce simultaneously to zero. The constants K and d depend upon the nature
of the liquid. K assumes numerical values between 2 and 2.23 erg/
mole$^{2/3}$, whereas d ranges between 6° and 9°C. For sufficiently
low temperatures and whenever a great accuracy is not desired, the
law of Ramsay and Shields can be replaced by Eötvos' formula,

$$\sigma = (K/\omega^{2/3}) \ (T - T_*)$$

In the latter formula, the numerical value of K is regarded as a constant independent of the nature of the liquid having a numerical
value of 2.22 erg/mole$^{2/3}$.

Surface Viscosity

So far, only strictly equilibrium situations have been considered. The possibility that the surface forces in a liquid in motion are different from those prevailing under hydrostatic conditions was advanced in 1913 by Boussinesq. According to Boussinesq's theory[5] the surface density of the force acting upon a free boundary of a liquid is compounded of the conventional normal surface tension of the surface and a tangential surface stress linearly dependent upon the components of the rate of surface strain according to the formula

$$\sigma_1 = \sigma + 2\eta_* \, \dot{e}_1 + \lambda_* \, (\dot{e}_1 + \dot{e}_2)$$

$$\sigma_2 = \sigma + 2\eta_* \, \dot{e}_2 + \lambda_* \, (\dot{e}_1 + \dot{e}_2)$$

where $\dot{e}_1$, $\dot{e}_2$ indicate respectively the principal components of the surface stress and rate of strain, and η_*, λ_* are coefficients of proportionality.

Boussinesq's theory of superficial viscosity was introduced on a strictly formal phenomenological basis and is obviously modeled upon the usual definition of volume viscosity. The physical reality of this phenomenon could be established either by experimentation or by an analysis of the transport of momentum by molecules moving in the field of force prevailing on or near the surface. Such a theory possibly could be constructed by extending Eyring's theory of viscosity in a liquid (see Ref. 6 Ch. IX) to some field of molecular force which can be expected to prevail in the vicinity of liquid surfaces.*

Inception of Capillary Phenomena†

After these preliminaries, let us proceed to the question of the determination of the range of gravitational or equivalently inertial accelerations for which capillary phenomena become important. For

---

*Eyring's theory of viscosity is based on the disparity between the probability of transition of a molecule from one local minimum of potential energy to an adjacent minimum, when a shear force is superimposed upon the more or less periodic molecular field of force. Surface viscosity effects could possibly be predicted on the basis of the difference between surface and interior molecular fields of force.

†Part of the discussion and results appearing in the present and the following two articles were taken from a previous publication[7] of the author.

this purpose, consider a liquid with one or more free surfaces and subjected to an acceleration ng, $g = 980.665$ cm/sec^2, denoting the standard average terrestrial acceleration of gravity. We shall further indicate with L the order of magnitude of the linear dimensions of the region of space of volume V occupied by the liquid. Whenever necessary, the more precise significance

$$L = 3\sqrt{V/\left(\frac{4\pi}{3}\right)} \tag{1}$$

can be attached to L. Thus defined, L is identical to the radius of the spherical configuration which the isolated liquid mass would assume under the sole action of surface tension forces.

The intensity of the capillary forces* acting upon the liquid will be of the order of σL, whereas the inertial forces acting upon the liquid will have the order of magnitude $ng\rho L^3$. The (nondimensional) ratio

$$n = \sigma/\rho ngL^2$$

of the above two expressions measures the relative importance of the capillary forces in an accelerated liquid. It can therefore be expected that capillary phenomena will become significant when $n \gg 1$, i.e., for liquids subjected to acceleration, corresponding to values of n smaller than

$$n_* = \sigma/\rho gL^2 \tag{2}$$

Numerical values of $n_* L^2$ for various typical liquids are given in Table 2. The estimates provided by the above considerations should, of course, find experimental verification. A confirmation of their plausibility can, however, be reached from the examination of situations amenable to rigorous analytical treatment. For this purpose we shall examine the static behavior of a liquid placed in a tank (see Figures 5a and b) of rectangular cross section, with one

---

*We shall denote with this term, indifferently, the cohesive (or surface tension) as well as the adhesive (or interfacial tension) forces. For the purpose of obtaining preliminary estimates, distinction between these forces (all of which originate from intermolecular actions and are accordingly of the same order of magnitude) appears unnecessary.

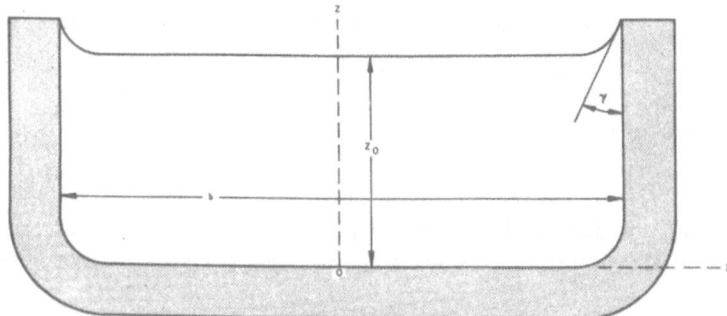

Figure 5a. Equilibrium of liquid in rectangular tank under action of gravity, surface and interfacial tension; adhesion predominant.

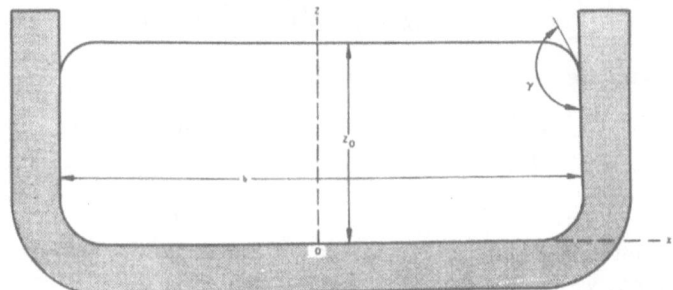

Figure 5b. Equilibrium of liquid in rectangular tank under action of gravity, surface and interfacial tension; cohesion predominant.

vertical side infinitely extended in a direction perpendicular to this cross section.*

We shall employ a set of cartesian coordinates x, z, having origin at the midpoint of the base, and z-axis vertical in a sense opposite to that of the applied acceleration. The shape of the vertical cross-section of the meniscus of the liquid can be described in terms of the system of nondimensionalized equations.†

---

*The terms "vertical" and "horizontal" are to be understood here as indicating directions respectively parallel and perpendicular to that of a (uniform) gravitational or inertial acceleration.

†The general problem of establishing the shape of a meniscus of a liquid in a tank consists of determining a free surface in such a manner that the total potential energy of the liquid—that is, the sum of its gravitational (or inertial) energy and its surface energy—be a minimum, subject to the restriction that the mass of the liquid remain constant. Details of the integration of the system of differential equations and the fulfillment of the boundary and normalization conditions appropriate to the situation described above can be found in Ref. 8.

$$\xi = K \{ 2 [\mathbf{D}(k) - D(k, \Phi)] - [\mathbf{K}(k) - F(k, \Phi)] \} \qquad (3)$$

$$\zeta = \{ \zeta_* \pm (2/k) \, \Delta(k, \Phi) \} \qquad (3')$$

in terms of the parameter

$$\Phi = (\pi - \alpha)/2 \qquad (4)$$

α being the inclination (with respect to the horizontal) of the surface of the liquid. In the above equations, $\xi = x/\ell$ $\zeta = z/\ell$ with

$$\ell = \sqrt{\sigma/\rho n g} \qquad (5)$$

Further,

$$\Delta(k, \Phi) = \sqrt{1 - k^2 \sin^2 \Phi}$$

The quantities

$$F(k, \Phi) = \int_0^\Phi d\phi/\Delta(k, \phi), \qquad D(k, \Phi) = \int_0^\Phi \sin^2 \phi \, d\phi/\Delta(k, \phi)$$

are elliptic integrals, the first being the Legendre integral of the first kind, whereas the latter can be obtained in the form

$$D(k, \Phi) = (1/k^2) \, [F(k, \Phi) - E(k, \Phi)]$$

in terms of $F(k, \Phi)$ and the elliptic integral

$$E(k, \Phi) = \int_0^\Phi \Delta(k, \phi) \, d\phi$$

of the second kind. $\mathbf{K}(k) = F(k, \pi/2)$ and $\mathbf{D}(k) = D(k, \pi/2)$ represent complete elliptic integrals. The parameter ζ_* is given by

$$\zeta_* = \overline{\zeta} - (\cos \gamma)/\xi \qquad (6)$$

where γ is the (constant) angle (of contact) at which the liquid intersects the vertical wall, and

$$\xi_1 = b/2 \, \ell, \qquad \overline{\zeta} = A/\ell^2 = \overline{z}/\ell \qquad (6')$$

b indicating the distance between the vertical sides of the tank, and $A = b\overline{z}$ the (constant) area of the intersection between the region of space occupied by the liquid and a vertical plane parallel to the z-axis.* A (or equivalently $\bar{z}$) is of course defined from the assumed knowledge of the amount of liquid present in a unit length of the tank. Finally the modulus k $(0 \leq k \leq 1)$ is defined by the boundary conditions $x(\Phi_1) = b/2$, i.e., on account of (3), by the transcendental equation

$$k\,[\,2D(k) - D(k,\,\Phi_1)] - [K(k) - F(k,\Phi_1)] = \xi_1 \qquad (7)$$

where in view of (4) and since the value a_1, which a assumes at the boundary of the tank is complementary to y,

$$\Phi_1 = (\pi - a_1)/2 = \pi/4 + y_1/2 \qquad (7')$$

As for the ambiguity of the second term of Eq. (3′) for ζ, the positive sign must be used when the adhesion between the liquid and the walls of the tank prevails over the cohesion, a concave meniscus thus being obtained; the negative sign leads to a convex meniscus corresponding to liquids in which cohesion forces are predominant.

For values of k sufficiently close to unity, the system of Eqs. (3), (3′) can be replaced by the approximate relation

$$\xi = \xi_1 \mp \left\{ \frac{1}{2} \log \frac{1 + \sqrt{1 - \left(\frac{\zeta_1 - \zeta_*}{2}\right)^2}}{1 - \sqrt{1 - \left(\frac{\zeta_1 - \zeta_*}{2}\right)^2}} \times \frac{1 - \sqrt{1 - \left(\frac{\zeta - \zeta_*}{2}\right)^2}}{1 + \sqrt{1 + \left(\frac{\zeta - \zeta_*}{2}\right)^2}} \right. $$
$$\left. + \, 2 \left[\sqrt{1 - \left(\frac{\zeta - \zeta_*}{2}\right)^2} - \sqrt{1 - \left(\frac{\zeta_1 - \zeta_*}{2}\right)^2} \right] \right\} \qquad (8)$$

between ζ and ξ, whereas for sufficiently small values of k, Eqs. (3), (3′) can be approximated by

$$\xi = (\xi_1/\cos y)\, \sin a \qquad (9)$$

---

*Equation (5) is a direct consequence of the normalization condition

$$\int_{-b}^{b} z\,dx = A \quad \text{or} \quad \int_{0}^{\xi_1} \zeta\,d\xi = 2\,\bar{\zeta}\,\xi_1$$

$$\zeta = \overline{\zeta} \mp \left[(\cos \gamma)/\zeta_1 + (\xi_1/\cos \gamma_1) \cos a \right] \qquad (9')$$

Equations (9), (9') are the parametric representation of a circle of radius $\xi_1/\cos \gamma$, having center at the point $\xi = 0$, $\zeta = \overline{\zeta} \mp (\cos \gamma)/\xi_1$.*

The previous procedure was applied to typical cases of water ($\rho = 1$ gr/cm^3, $\sigma = 72.75$ dyne/cm) and mercury ($\rho = 13.546$ gr/cm^3, $\sigma = 513.0$ dyne/cm) filling an infinite glass tank of rectangular cross section and vertical walls $b = 1$ in. $= 2.54$ cm apart up to a mean level of 0.75 in. $= 0.9525$ cm. The angle of contact between water and glass (whose value depends critically upon cleanliness and which apparently becomes zero for perfectly clean glass) was assumed to be $\gamma = 14$ deg. For mercury, a value of $\gamma = 40$ deg was taken. The determination of the value of k corresponding to a given load factor was obtained from the graphical representations (shown in Figures 6a and b) of the dependence of the function $\xi = \xi(k, \Phi_1)$ defined by Eq. (7) upon the value ($0 \leq k \leq 1$) of k, and for values $\Phi = 52$ deg, $\Phi_1 = 70$ deg respectively for water and mercury, of the angle Φ defined by Eq. (7'). The dependence of k upon n for these particular cases is shown graphically in Figures 6c and d. Both curves indicate that k assumes values nearly equal to 1 for n exceeding by a sufficient amount, a certain critical value n_*, and values very close to zero for values sufficiently smaller than n_*. The range of n (containing n_*) in which k varies between these two extremes is comparatively narrow. The critical load factor n_* can be defined as the value of n for which dk/dn is a maximum, i.e., the root of the equation

$$d^2 k/dn^2 = 0 \qquad (10)$$

---

*This conclusion is a special case of a more general result. As can can be easily demonstrated by the standard procedures of the calculus of variation, a liquid (whether free, or partially bounded by the walls of a container) will, under strictly zero gravity conditions, be bounded by a surface (or surfaces) of constant mean curvature. This geometrical condition, in conjunction with the boundary condition requiring the constancy of the angle of contact between the liquid and the walls of the container, determines the shape of the free surface. In particular, if the liquid is bounded by a vertical wall of cylindrical shape (devoid of sharp angles), its free surface will assume the shape of a (concave or convex) spherical cap. (Note that the previous results predict only the shape of possible figures of equilibrium, and do not make predictions regarding their stability.)

Writing Eq. (7) in the form

$$\xi_1 = \xi_1^o \sqrt{n} = \xi(k, \Phi_1)$$

where

$$\xi_1^o = b/2 \, \ell_0 \qquad\qquad (11)$$

with

$$\ell_0 = \sqrt{\sigma/\rho g} \qquad\qquad (11')$$

(assuming the value of 0.2724 cm for water, 0.1965 cm for mercury) we can put the relation between n and k in an explicit form. Let k_* be a root (in practice, the root) of k whose value is comprised between 0 and 1. The corresponding value of n will then be $n_* = [\xi(k_*, \Phi_1)/\xi_1^o]^2$. Substituting into this formula the definition Eq. (11), we obtain

$$n_* = \xi^2(k_*, \Phi_1) \, \sigma/\rho g \, (b/2)^2$$

Apart from the coefficient $\xi^2(k_*, \Phi_1)$, generally of the order of unity, and upon identifying $b/2$ with L, this exact formula is identical with Eq. (2). The approximate predictions of this latter expression appear thus to be verified, at least for this particular case. The actual shape of the menisci for liquid systems of the type contemplated here subjected to various load factors are shown for water and mercury in Figures 7a and b. It should be noted that the results just obtained (as well as others, which might be derived along similar lines in the future) could be used to predict the succession of configurations of the surface of a liquid under quasistatic conditions, i.e., when the intensity of the gravitational or inertial field changes with sufficient slowness so that the kinetic energy of the ensuing motion can be neglected, as compared with the potential energy of the liquid.

Time Scale of Motions of Liquids Under Action of Surface Forces

The next question to be answered regards the rate at which liquids move under the action of surface forces. Of particular interest is the time required by a liquid mass to modify its configuration from the one in which gravitational or inertial forces prevail to those in which surface tensions predominate. For this purpose let us consider a liquid that is neither accelerated nor under the action

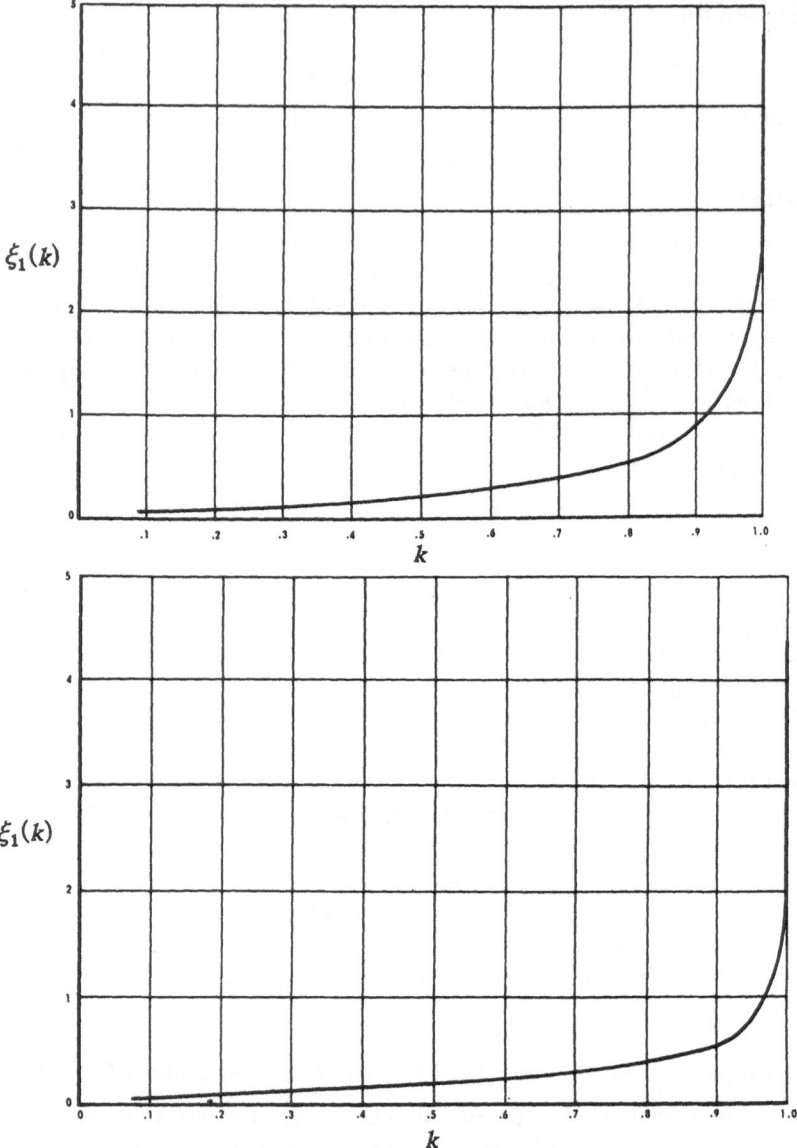

Figure 6a and b. Determination of shape of meniscus: graphical representation of function $\xi_1(k)$ for the determination of
modulus k, identifying shape of meniscus for water—
angle of contact 14 deg—(above), and mercury—angle of
of contact 40 deg (below).

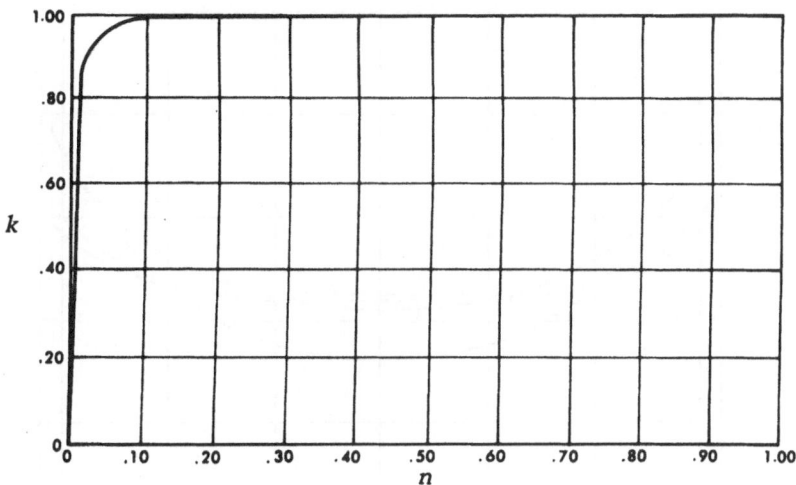

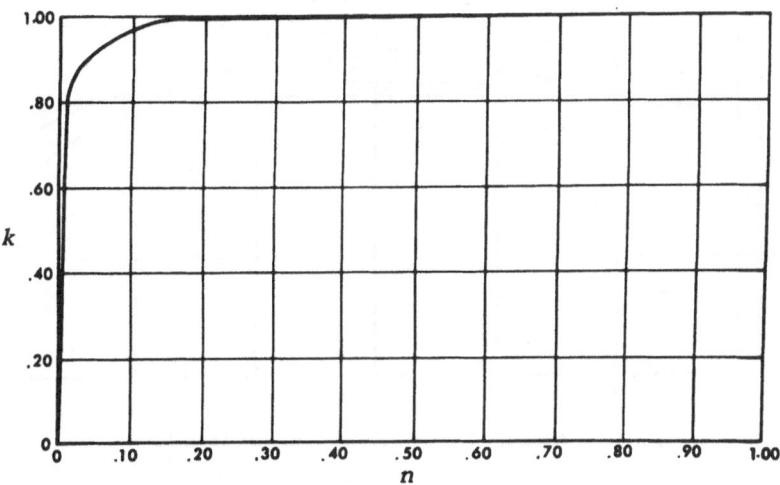

Figure 6c and d. Graphical representation of dependence of modulus k
upon load factor for water—angle of contact 40 deg—
(above), and mercury—angle of contact 40 deg (below).
Note abrupt change of modulus from a value $k \simeq 1$
identifying a largely horizontal surface, to a small
value of k, corresponding to a curved surface.

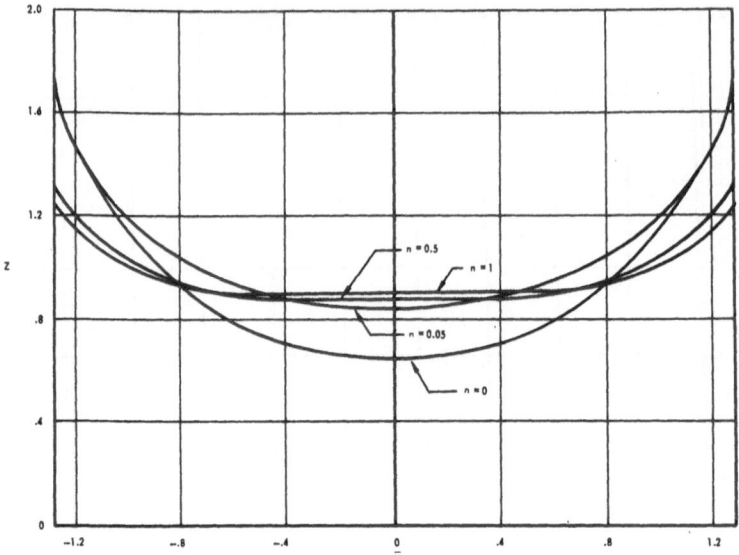

Figure 7a. Menisci for water for various load factors.

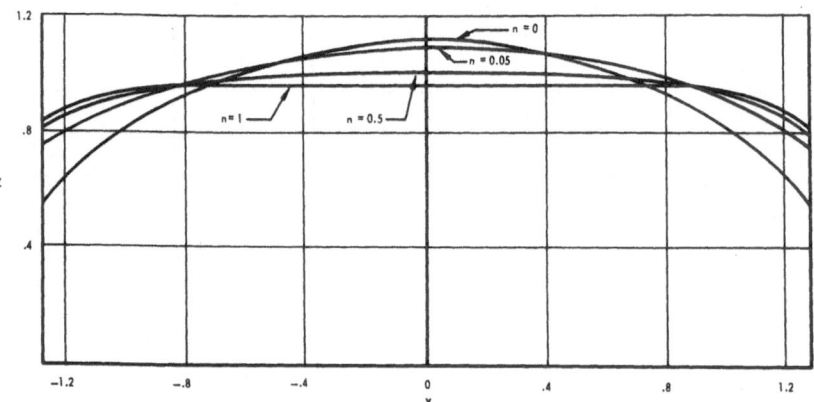

Figure 7b. Menisci for mercury for various load factors.

of any gravitational field. If the shape of the (free) surface of the liquid does not initially correspond to the equilibrium configuration—i.e., to a minimum of potential surface energy—the combined action of the surface tension and adhesion will cause the liquid to move, this motion being resisted by the inertia of the liquid mass.

Indicating as before with L the linear dimensions of the liquid body, its mass will be of the order of ρL^3. Since the forces causing its acceleration are of the order of σL, the deformation will proceed with an acceleration

$$a = \sigma/\rho L^2 \tag{12}$$

If the discrepancy* of the initial shape of the surface from that of the equilibrium configuration is also assumed to be of the order of L, the time required for the deformation to take place will be of the order of

$$t_* \simeq \sqrt{L/a} = \sqrt{\rho L^3/\sigma} \tag{13}$$

Numerical values of the parameter $t_*/L^{3/2}$ (that is, the value which t_* assumes for liquid systems having linear dimensions of the order of 1 cm) are given in Column 7 of Table 2.

The estimates afforded by Eq. (13) should, as in the previous case, be confirmed both experimentally and theoretically. The latter type of confirmation could be obtained, as before, from the knowledge of rigorous integrals for some particular cases of motion of a liquid under the preponderant action of surface forces. Available results of this type have been so far limited to the case of capillary waves on the surfaces of extended liquids, narrow jets and small drops.

A formal method for the general treatment of the dynamics of liquids either partially or totally bounded by free surfaces is given in Ref. 9. In all situations of interest, this method is based on the fact that the mass of the liquid is finite, i.e., it occupies a finite region of space. It is accordingly possible to define the configuration of the liquid in terms of a countably infinite† or, in particular, a finite number of parameters. Thus, for example, the height of each point of the free surface of a mass of liquid otherwise limited by the walls of a container with horizontal bottom and vertical walls can, in the cases of practical interest, be expressed in forms of an appropriate set of orthogonal functions. The coefficients of such an expansion would constitute then a set of parameters of the above type.

*This discrepancy can be defined, e.g., by the difference between the maximum dimensions of the initial configuration and a dimension of the configuration of equilibrium.

†A set of objects (in our case the parameters defining the configuration of the liquid) is said to be countably infinite whenever it can be "counted" i.e., wherever a one-to-one relation between the members of the set and the set of natural numbers 1,2,3... or 0, 1, 2... can be established.

TABLE 2: DATA REGARDING THE BEHAVIOR OF TYPICAL LIQUIDS UNDER ACTION OF SURFACE TENSION FORCES.

Liquid	Temp., °C	Adjacent medium	Surface tension, σ dyne/cm	Specific surface tension, σ/ρ cm³/sec²	[Max load factor]$\times L^2$ $\sigma/\rho g \times 10^2$ cm²	[Time Scale] $\times L^{3/2}$ $1/\sqrt{\sigma/\rho}$ sec	[Visco-capillary radius] $\frac{\nu^2}{\sigma/\rho}\times 10^6$ cm	Therm-cond. transition load factor $(1/c^4)(\nu_* / \nu_\rho \alpha)\times 10^2$ cm/°C
Water	20	Vapor of Substance	72.75	72.85	7.418	0.1172	1.381	1.31
Mercury	0		513.0	37.87	3.860	0.1625	4.061×10^{-2}	1.07×10
Ethyl Alcohol	20		22.30	28.25	2.881	0.1881	8.102	5.89
Acetone	20		23.70	29.92	3.051	0.1828	5.662×10^{-1}	0.13
Glycerine	20		63.40	49.84	5.080	0.1416	2.785×10^6	6.06×10^5
Olive oil	18	Air	33.10	36.37	3.710	0.1658	2.895×10^4	0.5
Liquid Zinc	360		877.0	135	13.770	0.0861	3.560×10^{-2}	2.1
Liquid hydrogen	−252		1.912	26.97	2.750	0.1926	8.287×10^{-2}	—

The potential (if the motion is assumed to be irrotational*) and the kinetic energy of the liquid can then be expressed in terms of the above parameters, functioning as (generalized) dynamical coordinates and hence the differential equations of motion can be set up in the conventional Lagrangian form. The problem of predicting the motion given the initial shape of the surface of the liquid and the distribution of velocity can thus be reduced to the integration of a Poisson (linear, partial differential) equation for the determination of the velocity potential, and the integration of the system of nonlinear Lagrangian differential equations of motion. The latter task is generally prohibitive, with the exception of some cases in which the deviation of the coordinates about their equilibrium value can be regarded as small. Such (linearizable) problems correspond to the situations already mentioned in which the motion of the liquid can be described in terms of superposition of capillary waves. With few exceptions, the limitations of the practical application of the above procedure to cases in which linearization of the Lagrangian equations is legitimate is, as to be expected, typical of all hydrodynamic problems involving free liquid surfaces. However, the value of these procedures should reside (in common with other developments in analytical dynamics) in affording an insight into the mathematical nature of the problem, and in providing useful results of general nature—in particular, in approaching the important problem of the kinematic stability of motion of liquids controlled by surface forces.

In some cases of particular simplicity, exact solutions can be obtained by means of the above method. For the time being we shall present, as a simplified example of application of the procedures involved, the treatment of the motion of a spheroidal mass of liquid under the sole action of surface tension forces[10]. (Aside from the reason of simplicity, this example was chosen because it reflects a situation amenable to experimental verification in an artificially produced zero gravity environment.) Let us consider a liquid spheroid having semiaxis (of symmetry) a, the remaining semiaxes being of length b. We shall limit our considerations to motions for which the spheroidal shape of the liquid is preserved.

---

*This specialization could be justified by the fact that experimental investigations could be restricted to liquids initially at rest. At any rate, the theoretical developments of Ref. 9 could presumably be extended in fairly direct fashion to rotational motions by the introduction of additional degrees of freedom.

The components of the velocity of the (only possible) irrotational motion will then be

$$v_x = (\dot{a}/a)\dot{x},\, v_y = (\dot{b}/b)\,\dot{y},\; v_z = (\dot{b}/b)\,\dot{z}$$

with respect to a system of cartesian coordinates having origin at the center of the spheroid and x-y plane coincident with the equatorial plane of the latter. The kinetic energy of the liquid will be given by

$$\mathfrak{I} = (1/2)\,\rho \int (v_x{}^2 + v_y{}^2 + v_z{}^2)\, d\mathcal{U} = (1/10)\, M (\dot{a}^2 + 2\dot{b}^2) \quad (14)$$

where $\mathcal{U}$ indicates the volume, M the mass of the spheroid. The potential energy of the liquid will be given by

$$U = \sigma A \tag{15}$$

The area A of the surface bounding the liquid can be expressed in the equivalent forms

$$
\begin{aligned}
A &= 2\pi\, [b^2 + (ab/\epsilon)\, \sin^{-1}\epsilon] \\
&= 2\pi\, [b^2 + (a^2/\epsilon')\, \sinh^{-1}(b/a)\epsilon']
\end{aligned}
\tag{16}
$$

where we have denoted with

$$\epsilon = \sqrt{1 - (b/a)^2}, \qquad \epsilon' = \sqrt{1 - (a/b)^2}$$

the possible expressions of the real eccentricity of the axial cross section of the spheroid. The constraint represented by the requirement that the volume $(4\pi/3)ab^2$ of the liquid (supposed incompressible) remain constant, provides the possibility of describing the configuration of the liquid in terms of a single coordinate; as such we can conveniently choose the parameter

$$\xi = L/a \tag{17}$$

where, consistent with the general definition,

$$L = \sqrt[3]{a^2 b}$$

represents the radius of the sphere having the same volume as the spheroid.

Substituting Eq. (17) into Eqs. (14) (15), it is further convenient to replace the time t by the nondimensional variable

$$\tau = t/t_*$$

with

$$t_* = \sqrt{ML^2/\sigma A_0} = \sqrt{(1/15)\frac{L^2}{\sigma/\rho}} \qquad (18)$$

M indicating the mass of the sphere of radius L, and A its area. The latter corresponds, incidentally, to the minimum area achievable by a set of equivoluminar spheroids.

Substituting Eqs. (17), (18) into Eqs. (14), (15) and (16), it follows that the total energy $E = \mathfrak{I} + U$ of the liquid system can be put in the form

$$E = (1/2)\sigma A_0 \{[(1 + \tfrac{1}{2} \xi^3)/\xi^4] (d\xi/d\tau)^2 + f(\xi)\} \qquad (19)$$

with

$$f(\xi) = \xi + \left(\sin^{-1}\sqrt{1 - \xi^3}\right)\Big/\sqrt{\xi(1 - \xi^3)}$$

$$= \xi + \left(\sinh^{-1}\sqrt{\xi^3 - 1}\right)\Big/\sqrt{\xi(\xi^3 - 1)} \qquad (20)$$

The motion of the (conservative) liquid system must occur in such a manner as to preserve the initial value $E = (\tfrac{1}{2})\sigma A_0 h$ of Eq. (19). From this identity we immediately obtain the integral

$$t = \pm t_* \int_{L/a_0}^{L/a} \sqrt{(1 + \tfrac{1}{2} \xi^3)/[h - f(\xi)]}\, d\xi/\xi^2 \qquad (21)$$

providing the time required by the spheroid to effect the transition between two configurations defined by length a_0 and a of the semiaxis of rotation. The ambiguity in sign of Eq. (21) is easily disposed of from a knowledge of the direction of the process. In using Eq. (21), care should be taken to account for the fact that the integrand is periodic, with period

$$T = \left| 2t_* \int_{L/a_0}^{L/a_1} \sqrt{(1 - \tfrac{1}{2} \xi^3)/[h - f(\xi)]}\, d\xi/\xi^2 \right| \qquad (22)$$

where $a_0 = L\xi_0$, $a_1 = L\xi_1$, represent the minimum and maximum value of a. Insomuch as these values are achieved when the kinetic energy of the system is zero, $\xi_0 \leq 1$, $\xi_1 \geq 1$ can be obtained, in view of Eq. (19), as the roots of the equation $f(\xi) = h$, $f(\xi)$ being defined by Eq. (20).

The deformation of a liquid spheroid starting from rest from an initial elongation defined by $a/L = 5$ has been computed on the basis of Eq. (21). The results are represented graphically in Figure 8 in nondimensional form for one full period. As already indicated, the above developments describe a possible liquid motion.

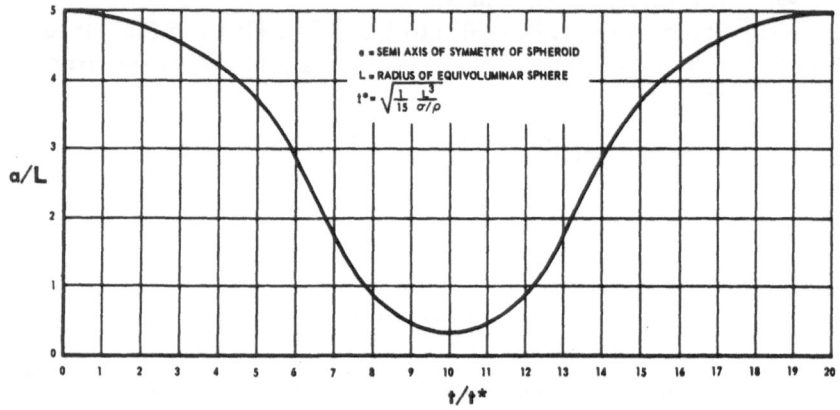

Figure 8. Deformation of liquid spheroid under action of surface tension.

In order to predict the possibility of its actual occurrence, the kinematic stability of the spheroidal shape should be investigated. Such an analysis was not felt necessary for the present purposes; however, drops of mercury have been observed to oscillate about a spherical shape in a zero gravity environment, preserving an apparent ellipsoidal shape in the process. Apart from these considerations, the critical time t_* defined by Eq. (18) differs from the estimated value by a factor $1/\sqrt{15} = 0.2582$. It might be more significant to identify the half period of oscillation given by Eq. (22) with t_*; however, this could depend upon the peculiarities of the particular motion through the integration constant h. In the available numerical example this quantity turns out to be $T = 2.594$. In any event, the predictions of Eq. (13) appear to be valid for purposes of estimate.

Influence of Viscosity

The fact that the retarding effects of viscosity have been neglected in the developments of the preceding article merits a few words of justification. The order of magnitude of the speed of flow of a liquid under the action of surface tension forces will be given by $v \simeq \sqrt{aL}$ so that we obtain from Eq. (12)

$$v \simeq \sqrt{\sigma / \rho L} \qquad (23)$$

indicating with ν the kinematic viscosity of the liquid, the density of the rate of dissipation of energy due to viscous friction will be given by

$$\dot{w} = \rho \nu (v/L)^2 \qquad (24)$$

so that for the total rate of dissipation of energy we shall have

$$\dot{W} = \dot{w} L^3 = \nu \rho v^2 L \simeq \nu \sigma \qquad (25)$$

where the estimate of v provided, for motion of the type considered here, by Eq. (23) was employed.

The rate at which the surface tension forces perform work can be estimated by the expression

$$\dot{U} = d(\sigma L^2) / dt \simeq \sigma L v \qquad (26)$$

That is, again using Eq. (23)

$$\dot{U} \simeq \sqrt{\sigma^3 L / \rho}$$

From Eqs. (23), (25) we obtain the estimate

$$Q = \dot{U} / \dot{W} \simeq \sqrt{\sigma L / \rho \nu^2} \qquad (27)$$

for the relative importance of capillary and viscous processes. Introducing the length

$$r_* = \rho \nu^2 / \sigma \qquad (28)$$

which could be conveniently termed the "viscocapillary radius" of the liquid, Eq. (26) can be also expressed in the more concise form

$$Q \simeq \sqrt{L / r_*} \qquad (27')$$

Numerical values of r_* for typical liquids are supplied in Column 8

of Table 2.   From the smallness (10^{-8} to 10^{-2} cm) of the values
there displayed*, it might be inferred that viscosity can be disre-
garded in dealing with motions initiated under the action of surface
tension forces in liquid samples of microscopic dimensions.

Equation (27) can also be given another interpretation, which
might promote an understanding of its significance by casting it in
the equivalent form

$$Q = \tau_* / t_* \qquad\qquad (27'')$$

where t_* is defined by Eq. (13), whereas

$$\tau_* = L^2 / \nu \qquad\qquad (28)$$

is well known as a measure of the relaxation time for liquid motion
decelerated under the action of viscous dissipative forces. It should
be observed that in equations such as (24) and (28), definition (1)
cannot always be regarded as entirely appropriate. The accuracy
of the estimates of these equations could be improved by reducing L
by a numerical factor, generally less than one and frequently of the
order of 1/10.   More accurate selection of this factor depends of
course upon a more exact knowledge of the geometry of the situation.

Heat Transfer—Transition Between Convective and Conductive Transfer Mechanism

The action of a gravitational and inertial field is necessary for
the propagation of heat by the so-called process of free convection
to take place in a fluid. Consequently, in a fluid at rest in an environment in which such forces are absent, heat can be transferred
only by its much less efficient (molecular) mechanism of conduction.   Conduction by forced convection can of course be achieved
in the absence of gravity in fluids maintained in a state of motion
by appropriate means. If the fluid is at some initial instant in motion, and if the primary cause of this motion (such as an applied
force—possibly of gravitational or inertial nature, moving solid
boundaries or an unsteady temperature difference) ceases to act, the
initially present fluid currents will decay at a rate estimated by Eq.
(28).   The amount of heat transferred by convection will decrease

---

*With the exception of Glycerol.

concomitantly. Eventually, although after a time possibly much larger than the one estimated by Eq. (28), this form of transfer will become insignificant. Less obvious is the phenomenon of heat transfer in a fluid subject to a very low gravitational or inertial field. Before proceeding with a discussion of such processes, let us consider the so-called Grashof number

$$\eta_G = L^3 \, nga\Delta T / \nu^2 \tag{29}$$

whose well-known function is to furnish an estimate of the importance of the force of buoyancy caused by a temperature difference ΔT in a heavy or accelerated fluid of expansion coefficient a, as compared to the retarding effect of viscous forces, the intensity of the free convection currents being determined by the balance between these two actions. Employing definitions in Eqs. (3) and (28), Eq. (29) can be put in the form

$$\eta_G(n) = (n/n_*)(a\Delta T / r_*) \tag{29'}$$

more directly useful for our purposes.

Consider the Nusselt number

$$\eta_N = hL/k$$

which provides an approximate determination of the importance of the process of heat transfer by convection as compared with that by conduction in a fluid occupying a region of space of linear dimensions L and for which the coefficients of convective and conductive heat transfer are respectively h and k. Introducing further the Prandtl number

$$\eta_P = C_P \, \eta/\kappa$$

where C_P denotes the specific heat at constant pressure of the fluid and η its viscosity, the Nusselt number can be put in the form

$$\eta_N = C \sqrt[4]{\eta_P \eta_G} \tag{30}$$

C being a coefficient which, for a gas heated by a vertical wall maintained at a constant temperature is equal to 0.52. Using Eq. (29) for Grashof's number, it follows from Eq. (30) that Nusselt's number can be represented in the form

$$\eta_N(n) = \eta_G(1) \sqrt[4]{n} \qquad (31)$$

The transfer of heat by conduction and free convection can be expected to assume equal importance when η_N is of the order of unity. Because of Eqs. (31) and (29′) this will occur when the intensity of the applied gravitational or inertial field is defined by the load factor

$$N_* = (1/C^4)\,(1/\eta_P)\,(r_*/La\Delta T) \qquad (32)$$

Heat transfer by conduction will presumably predominate when $n < N_*$. Typical numerical values of the parameter $N_* L$ are shown in the last column of Table 2. The above considerations presuppose ideal situations in which vapor bubbles are present or generated within the liquid. In actual situations, particularly when large temperature differences are involved, such bubbles would, of course significantly modify the mechanism of heat transfer.

Electrocapillary Effects

In view of the fact that liquids might acquire an electrostatic charge in many situations of practical importance, an estimate of the possible effects of such a charge appears in order. For this purpose, let us consider that a liquid, over whose surface an electric charge q is spread, will acquire an electrostatic energy which can be written in the form

$$U^{(e)} = (1/2)\,q^2/C = (1/2)\,C\mathsf{U}^2 \qquad (33)$$

where C denotes the capacity of the liquid system and U the electrostatic potential to which its surface is raised. Insomuch as $C \simeq L$ (this relation becoming, incidentally, an equality in the case of liquid masses of spherical shape), it follows from Eq. (33) that the electrostatic energy will become comparable to its capillary energy $U \simeq \sigma L^2$ whenever the linear dimension of the liquid mass assumes a value

$$r_*^{(e)} \simeq \sqrt[3]{q^2/\sigma} \simeq \mathsf{U}^2/\sigma \qquad (34)$$

The above expressions can be written in the form

$$r_*^{(e)} \simeq \mathsf{U}^2/9 \times 10^4\,\sigma \ \text{cm} \qquad (34')$$

if U is expressed in volts, σ in dyne/cm.

Insomuch as the surface energy is proportional to L^2, it increases more rapidly with L than either of the expressions in Eq. (33), one which increases only linearly with L, the other decreasing as L^{-2}. Therefore, the effects of an electrostatic charge will predominate for liquid masses having size smaller than $r_*^{(e)}$. Assuming σ to be of the order of 50 dyne/cm, it follows from Eq. (34') that

$$r_* \simeq 2 \times 10^{-7} \ cm \ (\mathbb{U} \ in \ volt)$$

Thus, if the electrostatic potential of the liquid is taken to be of the order of 1 volt, electrostatic effects are of importance only in the case of microscopic droplets, whereas in order to obtain noticeable effects with macroscopic ($r_* \simeq 1$ cm) samples of liquid, potentials of the order of 2000 volts would be required.

Barocapillary Effects

The smallness of the intermolecular forces responsible for the surface tension phenomena suggests the possibility that, in the cases of liquids upon which no external gravitational fields are acting, the gravitational field of the liquid mass itself (i.e., the intermolecular gravitational attraction) might acquire some importance.

We shall investigate this possibility by limiting our considerations to the case of liquid spherical masses. A liquid sphere of radius R isolated in space will possess a gravitational energy

$$U^{(g)} = - (8\pi^2/15) \ G \rho^2 R^5$$

where $G = 0.6668 \times 10^{-7}$ dyne cm^2/gr, indicates the gravitational constant. This amount of energy will have a value comparable to the capillary energy $4\pi\sigma R^2$ of the sphere whenever the radius of this sphere is of the order of what might be called the "barocapillary radius"

$$r_*^{(g)} \simeq \sqrt[3]{\sigma/G\rho^2} \tag{35}$$

of the liquid. Assuming, for the purposes of a crude estimate, $\rho \simeq 1$ gr/cm^2, $\sigma \simeq 50$ dyne/cm^2, we obtain $r_* \simeq 20$ m. Self-gravitational effects would thus be expected to become significant for liquid masses of rather large, but not immense, dimensions.

References

1. H. Minkowski, "Kapillarität," in *Encyklopädie der Mathematischen Wissenschaften*, Vol. V, Part 2, Leipzig, 1903-1921, pp. 594-602.

2. A. Harasima, "Molecular Theory of Surface Tension," in *Adv. in Chem. Phys.*, Vol. 1, Interscience, New York, 1959, pp. 203-237.

3. C. V. Raman and L. A. Randas, "On the Thickness of the Optical Transition Layer in Liquid Surfaces," *Philos. Magazine*, 3 1927, p. 220.

4. M. W. Zemanski, *Heat and Thermodynamics*, McGraw-Hill, New York, 1943.

5. J. Boussinesq, "Sur l'Existence d'une Viscosité Superficielle, dans la Mince Couche de Transition d'une liquide d'un autre fluide contigu," *Compt. Rend. de l'Acad. des Sciences*, 156 1913, p. 983.

6. S. Glasstone, K. J. Laidler and H. Eyring, *Theory of Rate Processes*, McGraw-Hill, New York, 1961.

7. E. T. Benedikt, "Scale of Separation Phenomena in Liquids Under Conditions of Nearly Free Fall," *J. Amer. Rocket Soc.*, 29, 1959, p. 150.

8. E. T. Benedikt, "Epihydrostatics of a Liquid in a Rectangular Tank with Vertical Walls," ASRL–TM–60–38, Norair Division, Northrop Corporation, Hawthorne, Calif.

9. E. T. Benedikt, "Epihydrodynamics: The Dynamics of Liquids under the Action of Surface Tension Forces," presented at The Divisional Meeting of The Division of Fluid Dynamics, Amer. Phys. Soc., University of Michigan, Ann Arbor, Mich., 1959.

10. E. T. Benedikt, "Dynamics of a Liquid Spheroid under The Action of Cohesive Forces," ASRL–TM–60–10, Norair Division, Northrop Corporation, Hawthorne, Calif.

ON THE MATHEMATICAL ASPECTS
OF BENEDIKT'S REPRESENTATION
OF THE EQUATIONS OF EPIHYDRODYNAMICS

M. Z. v. KRZYWOBLOCKI

Michigan State University, East Lansing, Michigan

Benedikt[2] assumes the following configuration: the liquid systems occupy a finite region of space. The geometry of the surfaces bounding the region in question can be identified by some number of parameters, which can be used as generalized coordinates. The dynamics of the liquid system is treated by the Lagrangian method. Application of this method requires the knowledge of the potential energy in the case under consideration, of the surface (free) energy, proportional to the area of the surfaces bounding the liquid, of the gravitational energy as well as of the kinetic energy. In order to obtain the latter, the potential of the flow, which is regarded as laminar, can be represented as a linear function of the generalized velocities. The coefficients of this expression are certain functions of the configuration which can in principle be determined by the standard procedures of potential theory.

An expansion of the kinetic energy as a quadratic function of the generalized velocities is derived in Ref. 2 in terms of these coefficients. A Langrangian function is subsequently set up by combining the above result with expressions of potential (surface and gravitational) energy. The main characteristics of the liquid in question such as forces acting upon the system, surface tension effects, pressure distribution, etc., are also formulated in terms of generalized coordinates and velocities in Ref. 2.*

Characteristic Properties of the Configuration Space

The instantaneous configuration of a system is described by means of the values of "n" generalized coordinates, $q_1, \ldots, q_n$, and corresponds to a particular point in a Cartesian hyperspace where the q's form the n-coordinate axes. This n dimensional space is called a configuration space. As time goes on, the state of the system changes, and the system point moves in configuration space

---

*Discussion of some aspects of this method can be found also in Article 6 of the preceding contribution. Additional work in this field is presently in progress, and it is hoped that new results will be available in the open literature in the near future.

tracing out a curve, known as the path of motion of the system. The motion of the system refers to the motion of the system point along the path in the configuration space. Time can be considered formally as a parameter of the curve; to each point on the path there is associated one or more values of the time. Each point on the path in the configuration space represents the entire system configuration at some given instant of time.

The formalism of classical mechanics, as employed and presented above, refers to systems with a finite or at most a denumerably infinite number of degrees of freedom. A material system is regarded from the dynamical point of view as constituted of a number of particles, subject to interconnections and constraints of various kinds. The systems considered are such that the number of coordinates, n, is finite. It is obvious that from a practical standpoint this may be not the most convenient formalism, particularly when applied to continuous media like fluids. In this case one cannot assume that the medium in question consists of a finite number of particles, however large. Obviously, it seems to be more adequate to consider the medium in question as a continuum; but in this case the methods of classical mechanics cannot be used. We propose below a different approach to this problem.

The Lagrangian Formulation for Continuous Systems and Fields

Assume a continuous system described by means of three coordinates and the time $\{x_1,\ x_2,\ x_3;\ t\}$. Let us denote the displacements of a particle from its equilibrium position by $\eta_j(x_1,\ x_2,\ x_3;\ t)$, $(j = 1, 2, \ldots)$, and let us introduce the following:

Lagrangian:

$$L = \iiint \mathcal{L}\, dxdydz \tag{1}$$

Lagrangian density:

$$\mathcal{L} = \mathcal{L}\ (\eta_{j,x};\ \eta_{j,y};\ \eta_{j,z};\ \eta_{j,t}\ x,\ y,\ z;\ t\ ); \tag{2}$$

Hamilton's principle:

$$\delta I = \delta \int_1^2 \iiint \mathcal{L}\ dxdydzdt = 0 \tag{3}$$

Equation of motion for each η_j of the form:

$$\frac{d}{dt}\frac{\partial \mathcal{L}}{\partial \dot{\eta}_j} + \sum_k \frac{d}{dx_k}\left(\frac{\partial \mathcal{L}}{\partial(\partial\eta_j/\partial x_k)}\right) - \frac{\partial \mathcal{L}}{\partial \eta_j} = 0 \qquad (4)$$

$$j = 1, 2, \ldots$$

Functional or variational derivative of L:

$$\frac{\delta L}{\delta \eta_j} = \frac{\partial \mathcal{L}}{\partial \eta_j} - \sum_{k=1}^{3} \frac{d}{dx_k}\frac{\partial \mathcal{L}}{\partial(\partial\eta_j/\partial x_k)} \qquad (5)$$

$$\frac{\delta L}{\delta \dot{\eta}_j} = \frac{\partial \mathcal{L}}{\partial \dot{\eta}_j} \qquad (6)$$

The δ—variation of L:

$$\delta L = \int \sum_j \left(\frac{\delta L}{\delta \eta_j}\,\delta\eta_j + \frac{\delta L}{\delta \dot{\eta}_j}\,\delta\dot{\eta}_j\right) dV \qquad (7)$$

In a discrete system, the problem is set up in the Lagrangian formulation by finding the kinetic and potential energies and writing the Lagrangian as the difference of these quantities. In some problems in the mechanics of continuous systems, we seek the Lagrangian as the volume integral of a density $\mathcal{L}$. The kinetic and potential energies can similarly be obtained as volume integrals of densities $\mathfrak{J}$ and $\mathcal{U}$, respectively, with the relation:

$$\mathcal{L} = \mathfrak{J} - \mathcal{U} \qquad (8)$$

where $\mathfrak{J}$ denotes the kinetic energy density per unit volume and $\mathcal{U}$ the potential energy density per unit volume.

One can obtain a Hamiltonian formulation for systems with a continuous set of coordinates. Conjugate to η_i there is a canonical momentum:

$$p_i = \frac{\partial L}{\partial \dot{\eta}_i} \qquad (9)$$

the Hamiltonian for the system is:

$$H \equiv \sum_i p_i \dot{\eta}_i - L; \quad H = \iiint dx_1 dx_2 dx_3 \left(\sum_k \frac{\partial \mathcal{L}}{\partial \dot{\eta}_k}\,\dot{\eta}_k - \mathcal{L}\right) \qquad (10)$$

Momentum density, π:

$$\pi_i = \frac{\partial \mathcal{L}}{\partial \dot{\eta}_i} \tag{11}$$

Equation (10) is in the form of a space integral over a Hamiltonian density $\mathcal{H}$ defined by

$$\mathcal{H} = \sum_k \pi_k \dot{\eta}_k - \mathcal{L} \tag{12}$$

Hamilton's canonical field equations for the continuous system are

$$\frac{\delta H}{\delta \eta_k} = -\dot{\pi}_k; \quad \frac{\delta H}{\delta \pi_k} = \dot{\eta}_k; \quad \frac{\partial \mathcal{H}}{\partial t} = -\frac{\partial \mathcal{L}}{\partial t} \tag{13}$$

or

$$\frac{\partial \mathcal{H}}{\partial \eta_k} - \sum_j \frac{d}{dx_j} \frac{\partial \mathcal{H}}{\partial (\partial \eta_k / \partial x_j)} = -\dot{\pi}_k; \quad \frac{\partial \mathcal{H}}{\partial \pi_k} = \dot{\eta}_k \tag{14}$$

The Hamiltonian density is equal to the sum of the kinetic and potential energy densities:

$$\mathcal{H} = \mathfrak{I} + \mathfrak{V} \tag{15}$$

A system of n discrete degrees of freedom has n Lagrange equations of motion; for the continuous system with an infinite number of degrees of freedom, Eq. (4) furnishes a separate equation of motion for each value of x_k. The continuous nature of the indices x_k appears in that Eq. (4) is a partial differential equation in the four variables (x_1, x_2, x_3; t), yielding η as

$$\eta (x_1, \ x_2, \ x_3; \ t)$$

Linear Functional and the Fundamentals of the Statistical Fluid Dynamics

The possibility of applying a classical field theory technique to the Benedikt formulation of the equations of motion of epihydrodynamics was outlined above. Lack of precise initial or boundary data may necessitate using statistical methods. Briefly, one can talk about the application of the method of statistical fluid dynamics (not yet constructed) to epihydrodynamics. In connection with that, we shall discuss briefly the fundamental notions which have to be employed in the formulation of statistical fluid dynamics. We shall begin with some definitions.

Definition I. A nonempty class R of sets is called a ring if $E \cup F$ and E-F belong to R whenever E and F do. The ring is called a G-ring if it contains the union of every countable collection of its members. A ring contains the empty set, and it contains the union and intersection of any finite number of its members.

Definition II. A measure is a function μ defined on a ring R, the values of μ being either real numbers or $+\infty$, subject to the conditions:

(a) $\mu(E) \geq 0$ if $E \in R$;

(b) $\mu(\phi) = 0$; ϕ = empty set;

(c) If $\{E_n\}$ is a sequence of pairwise disjoint members of R whose union is in R, then $\mu(\bigcup_{n=1}^{\infty} E_n) = \sum_{n=1}^{\infty} \mu(E_n)$ (with the usual conventions about $+\infty$ in relation to the series). As a consequence of the definition, a measure has also the following properties:

(d) If $E, F \in R$, and $E \subset F$, then $\mu(E) \leq \mu(F)$.

(e) In situation (d), if $\mu(E) < \infty$, then $\mu(F - E) = \mu(F) - \mu(E)$.

Definition III. If T is any nonempty set, if S is a G-ring of subsets of T such that T is the union of all members of S, and if μ is a measure defined on S, we call (T, S, μ) a measure space. The members of S are called measurable sets. The set T itself need not belong to S, and the complement of a measurable set need not be measurable.

A measure space determines a class of functions called integrable, and with each such function x there is associated a real number called its integral and denoted by

$$\int_T x \, d\mu \quad \text{or} \quad \int x \, d\mu$$

One may omit the symbol T on the integral sign if the context is such that there is no ambiguity. To define the class of integrable functions, we begin with the concept of a measurable function.

Definition IV. A set X with a family $\mathfrak{S}$ of subsets is called a topological space if $\mathfrak{S}$ satisfies the following conditions:

(a) The empty set ϕ and the whole space X belong to $\mathfrak{S}$;

(b) The union of any number of members of $\mathfrak{S}$ (even uncountably many) is again a member of $\mathfrak{S}$;

(c) The intersection of any finite number of members of $\mathfrak{S}$ is again a member of $\mathfrak{S}$. The family $\mathfrak{S}$ is called a topology for X.

Definition V. A topological space is called a T_1 - space if each set consisting of a single point is closed.

Definition VI. A topological space is called a Hausdorff space if for each pair of distinct points x_1, x_2, $x_1 \neq x_2$, there exist disjoint neighborhoods U_1, U_2 of x_1 and x_2, respectively, i.e., there exist open sets S_1, S_2 such that $x_1 \in S_1$, $x_2 \in S_2$, and $S_1 \cap S_2 = \phi$.

Definition VII. If C is the class of all compact subsets of a locally compact Hausdorff space, the elements of the smallest **G**-ring containing C are called the Borel sets of the space. This smallest **G**-ring is the intersection of all **G**-rings which contain C.

Consider functions x defined on T, with values $x(t)$ in the set consisting of the real numbers and the symbols $+ \infty$, $- \infty$. We define $N(x)$ as the set of $t \in T$ such that $x(t) \neq 0$.

Definition VIII. The function x is called measurable (with respect to S) if $N(x) \cap x^{-1}(M) \in S$ (when in $y = F(x)$, F has an inverse, $y = F(x)$ is equivalent to $x = F^{-1}(y)$) for each Borel set M of real numbers and if also the sets $\{t: x(t) = + \infty\}$, $\{t: x(t) = - \infty\}$ belong to S.

Definition IX. The function x is called simple if there exists a finite collection of pairwise disjoint sets E_1, . . . , E_n in S such that the value of x on E_k is a real constant a_k (not $\pm \infty$) and the value of x is 0 on $T - (E_1 \cup . . . \cup E_n)$. Then $x = a_1 x_{E(1)} + ... + a_n x_{E(n)}$. The simple functions form a real linear space.

Definition X. A simple function x is called integrable if $\mu[N(x)] < \infty$. If x is an integrable simple function with the distinct nonzero values a_1, . . . , a_n, assumed on sets E_1, . . . , E_n, respectively, the integral of x is defined to be:

$$\int x\, d\mu = a_1 \mu(E_1) + . . . + a_n \mu(E_n)$$

If $x(t) \equiv 0$, we define the integral of x to be 0.

We are now ready to put down briefly the fundamental aspects of statistical fluid dynamics. The primary characteristic features of the statistical mechanics of a conservative dynamical system having a finite number of degrees of freedom and satisfying the Hamilton-Jacobi equations are the following:

(a) definition of a phase-space Ω; every state of the system is characterized by a point $\omega \in \Omega$, whose $2k$-coordinates are the Lagrangian parameters q_l and the conjugate momenta p_l

(b) definition of a measure μ in Ω invariant under the transformation $T_t\omega$.

(c) the proof of the uniqueness theorem: starting at the time $t = 0$ from a given initial state $\omega = \omega_0$, all the subsequent or prior states described as the transformations of ω, $T_t\omega$, where t varies from $-\infty$ to $+\infty$, are perfectly well determined and describe in Ω a curve of trajectory $\Gamma(\omega)$

(d) proof of the ergodic theorem: the time average,

$$\bar{F}_T = \lim_{T\to\infty} (2T)^{-1} \int_{-T}^{+T} F[T_t\omega]\, dt$$

computed along a trajectory $\Gamma(\omega)$ exists for almost all ω and is equal to the statistical average, $F(\omega) = \int_\Omega F(\omega)\, d\mu$, if the transformation $T_t\omega$ of Ω into itself is metrically transitive.   The sequence of transformations is called metrically transitive if the only sets which are invariant are the sets of measure zero and their complements.   In constructing the statistical mechanics of continuous media, one must consider a system having an infinite number of particles and consequently an infinite number of degrees of freedom; thus, in place of a phase-space, one must take a function-space.

A few examples, solved today, referring to the statistical mechanics of continuous media, deal with simple one-dimensional continuous elements: vibrating string (J. Kampé de Fériet); vibrating beams, one-dimensional wave propagation in rods and bars, Schrödinger equation (M. Z. v. Krzywoblocki).   In all the cases the ergodic theorem was proved.

Particular Example

In the mathematical formulation of the Lagrangian in Sec. 3, the displacement of the i-th particle from its equilibrium position is denoted by η_i.   Benedikt[2] proposes a simple example of contraction (or expansion) of liquid masses occupying certain regions in space and having a boundary which is assumed to be permanently of the shape of a rotationally symmetrical ellipsoid, under the action of surface tension (or contraction) forces. The expressions for the kinetic energy and surface energy in the problem in question are given in Ref. 2 in terms of the dimensionless semimajor axis, a, and its derivative with respect to time t. The kinetic and potential energies refer to the total volumes of liquid spheroids. To obtain the Lagrangian density one has, in a first approximation, to divide the kinetic and potential energies by the volume, V, of the spheroid. The quantity V is assumed to be roughly constant. Using the relation in Eq. (8) and the values of the kinetic and potential energies one obtains:

$$\mathcal{L} = \mathcal{J} - \mathcal{U} = V^{-1} \{1/10 \ ML^2 \ [1 + (2a^3)^{-1}] \ \dot{a}^2$$

$$- 1/2 \ \sigma \ S_0 \ [a^2 + (a^{1/2} \ \sinh^{-1}((a^3 - 1)^{1/2})(a^3 - 1)^{-1/2})]\} \quad (16)$$

To obtain the equations of motion, one needs the following derivatives (see Eq.(4))with $\eta \equiv a$:

$$\frac{\partial \mathcal{L}}{\partial \dot{\eta}_i} = V^{-1} \{2/10 \ ML^2 \ [1 + (2a^3)^{-1}] \ \dot{a} \} \quad (17)$$

$$\frac{\partial \mathcal{L}}{\partial \eta_i} = V^{-1} \{ -3/20 \ ML^2 \ a^{-4} \ \dot{a}^2 + \ldots \} \quad (18)$$

With

$$d\mathcal{L} = V^{-1} \{[-3/20 \ ML^2 \ \dot{a}^2 \ a^{-4} + \ldots] \ da + \ldots \} \quad (19)$$

and

$$da = \frac{\partial a}{\partial x} \ dx + \frac{\partial a}{\partial t} \ dt, \qquad \text{since} \qquad a = a \ (x, \ t)$$

one gets

$$\frac{\partial \mathcal{L}}{\partial(\partial \eta_i / \partial x_k)} = \frac{\partial \mathcal{L}}{\partial(\partial a / \partial x)} = V^{-1}\{-3/20 \ ML^2 \ \dot{a}^2 \ a^{-4} + \ldots \} \quad (20)$$

etc. Inserting these derivatives into Eq. (4) furnishes an equation for $a = a(x, \ t)$, which, if solved, will furnish the sought function $a(x, \ t)$.

Discussion of Some Possible Methods of Solutions

The field equations (4) obtained above furnish the equations of motion for the system in question. Below, we shall collect some possible methods of solution of such a system of equations. But first, let us remodel the field equations. Consider a real field described by a single field variable ϕ, which is a function of the space-time variable x_μ. With:

$$I = \int d^4x \; \mathcal{L}(\phi, \partial\phi/\partial x_\mu); \quad \overline{\phi_\mu = \partial\phi/\partial x_\mu} \qquad (21)$$

one gets:

$$\delta I = \int_\Omega d^4x \left[\delta\phi(\vec{x}) \; \frac{\partial\mathcal{L}}{\partial\phi(\vec{x})} + \delta\phi_\mu(\vec{x}) \frac{\partial\mathcal{L}}{\partial\phi_\mu} \right] \qquad (22)$$

Application of the Gauss-Ostrogradski theorem to the second term furnishes the result:

$$\delta I = \int_\Omega d^4x \cdot \delta\phi \left[\frac{\partial\mathcal{L}}{\partial\phi} - \frac{\partial}{\partial x^\mu}\left(\frac{\partial\mathcal{L}}{\partial\phi_\mu}\right) \right] + \int_\Sigma d^3x \cdot \delta\phi \left(\frac{\partial\mathcal{L}}{\partial\phi_\mu}\right) n_\mu(\vec{x}) \quad (23)$$

where Σ denotes the surface of Ω and $n_\mu(\vec{x})$ is the outward normal to this surface at the point $\vec{x}$. With $\delta I = 0$ and $\delta\phi = 0$ on Σ, one gets the field equations in the form:

$$\frac{\partial\mathcal{L}}{\partial\phi} - \frac{\partial}{\partial x^\mu} \; \frac{\partial\mathcal{L}}{\partial\phi_\mu} = 0 \qquad (24)$$

In a continuous infinity of variables, the field is defined by an infinite number of coordinates $\phi(x, y, z, t)$, one for each point in space-time. Dividing the continuous three-dimensional space into small cells, Q_s, with $L = \sum_s \mathcal{L}_s \; \delta x_s$, the Lagrangian L of the whole system being a function of all the Q_s's and $\dot{Q}_s$'s, $\mathcal{L}_s$ being the Lagrangian density $\mathcal{L}$ in the cell "s", the action is given by the expression:

$$I = \int dt \cdot L(Q_1, Q_2, \ldots; \; \dot{Q}_1, \dot{Q}_2, \ldots) \rightarrow \int dt \cdot d^3x \, \mathcal{L}(\phi, \phi_\mu) \qquad (25)$$

A chosen method of solution of the system of field equations depends upon the form of $\mathcal{L}$. Some particular cases may be briefly discussed. (For more details see Refs. 3, 4).

Real one-component field. (Noninteracting neutral and complex scalar fields).

$$\mathcal{L} = -\frac{1}{2}\left(\mathcal{K}^2\phi^2 + g_{\mu\nu}\frac{\partial\phi}{\partial x_\mu}\frac{\partial\phi}{\partial x_\nu}\right)$$

$$= -\frac{1}{2}\left(\mathcal{K}^2\phi^2 + \frac{\partial\phi}{\partial x_\lambda}\frac{\partial\phi}{\partial x^\lambda}\right) \qquad (26)$$

$$\frac{\partial\mathcal{L}}{\partial\phi} = -\mathcal{K}^2\phi\; ;\quad \frac{\partial\mathcal{L}}{\partial\phi^\lambda} = \frac{\partial\mathcal{L}}{\partial(\partial\phi/\partial x_\lambda)} = \phi_\lambda \qquad (27)$$

The field equation is (with $x_0 \equiv t$):

$$\Box\phi - \mathcal{K}^2\phi = 0;\quad \Box = \frac{\partial^2}{\partial x_1{}^2} + \frac{\partial^2}{\partial x_2{}^2} + \frac{\partial^2}{\partial x_3{}^2} - \frac{\partial^2}{\partial x_0{}^2} \qquad (28)$$

The Klein-Gordon equation is of this type. The simplest solution of this equation is the plane wave

$$\phi(x) = a\ \exp\ (i\,k_\mu\,x_\mu);\quad k_\mu{}^2 = -\mathcal{K}^2 \qquad (29)$$

By means of this basic solution one may construct general solutions of Eq. (29) (for details see Ref. 4, p. 140) (four-dimensional Fourier integral):

$$\phi(x) = (2\pi)^{-3}\int \exp\ (i\,k_i x_i)\cdot(4\pi\lambda)^{-1}\ \{a\exp\ (-i\lambda x_0) \qquad (30)$$

$$+\ \bar{a}\ \exp\ (i\lambda x_0)\ \}\ \cdot\ ds_0$$

$$a = a(k_i,\ \lambda);\quad \bar{a} = \bar{a}\ (k_i,\ -\lambda);\quad ds_0 = dk_1 dk_2 dk_3 \qquad (31)$$

$$\lambda^2(k_i) = k_i{}^2 + \mathcal{K}^2,\ (i = 1,2,3);\quad k_0 = \pm\ (k_i{}^2 + \mathcal{K}^2)^{1/2} \qquad (32)$$

This three-dimensional Fourier representation of the solution contains two arbitrary functions a and $\bar{a}$ of the three variables k_1, k_2, k_3.

Another representation of the solution is by means of line integrals in the complex k_0-plane:

$$\phi(x) = (2\pi)^{-4} \oint \frac{a(k) \exp{(i\,k_\mu x_\mu)}}{(k_\mu{}^2 + \mathcal{K}^2)} \, dk \qquad (33)$$

$$a(k) = a(k_1, k_2, k_3, i\,k_0)$$

The integration over the variables k_1, k_2, k_3 is carried out from $-\infty$ to $+\infty$, the integration over k_0 is carried out in the complex k_0-plane, the closed path of integration encircling, in the positive sense, both poles $[k_0 = \pm \lambda(k_i)]$ of the integrand. The initial value problem has the following solution: any solution $\phi(x)$ of Eq. (28) at an arbitrary point x in the domain under consideration can be expressed by means of the initial values of $\phi(x)$ and the normal derivative $\partial_n \phi(x)$ at points on an arbitrary spacelike surface σ of this domain:

$$\phi(x) = \int_\sigma \Delta(x - x') X_\mu' \phi(x') d\sigma_\mu' \qquad (34)$$

where X_μ is the differential operator defined by the equation

$$X_\mu = \overrightarrow{\partial}_\mu - \overleftarrow{\partial}_\mu \qquad (35)$$

the arrows indicating the direction in which the differentiations are to be carried out. The Δ-function is the function ϕ in Eq. (33) with $a(k) \equiv 1$.

The technique outlined above can be generalized to a complex scalar field. With

$$W = \int_{\sigma_1}^{\sigma_2} \mathcal{L}^{(1)} dx; \quad \mathcal{L}^{(1)} = -(\psi_{,\nu}{}^* \psi_{,\nu} + \mathcal{K}^2 \psi^* \psi) \qquad (36)$$

$$\psi^* = \text{complex conjugate field},$$

The corresponding Euler-Lagrange equations are

$$(\square - \mathcal{K}^2)\, \psi = 0; \quad (\square - \mathcal{K}^2)\, \psi^* = 0 \qquad (37)$$

Noninteracting—the first rank bispinor fields. The action functional:

$$W = \int_{\sigma_1}^{\sigma_2} \mathcal{L}^{(1)} dx; \quad \mathcal{L}^{(1)} = -\frac{1}{2}(\bar{\psi}\gamma_\mu \psi_{,\mu} - \bar{\psi}_{,\mu}\gamma_\mu \psi) - \mathcal{K}\bar{\psi}\psi \qquad (38)$$

with $\bar{\psi} = \gamma_4 \psi^*$, leads to the Euler-Lagrange equations which are identical with the first order Dirac equations:

$$\gamma_\mu \psi_{,\mu} + \mathcal{K} \psi = 0 ; \quad \gamma_\mu \bar{\psi}_{,\mu} - \mathcal{K} \bar{\psi} = 0 \tag{39}$$

Here γ_μ are the well-known matrices,

$$\gamma_1 = \begin{pmatrix} 0 & 0 & 0 & i \\ 0 & 0 & i & 0 \\ 0 & -i & 0 & 0 \\ -i & 0 & 0 & 0 \end{pmatrix} \quad \gamma_2 = \begin{pmatrix} 0 & 0 & 0 & 1 \\ 0 & 0 & -1 & 0 \\ 0 & -1 & 0 & 0 \\ 1 & 0 & 0 & 0 \end{pmatrix}$$

$$\tag{40}$$

$$\gamma_3 = \begin{pmatrix} 0 & 0 & i & 0 \\ 0 & 0 & 0 & -i \\ -i & 0 & 0 & 0 \\ 0 & i & 0 & 0 \end{pmatrix} \quad \gamma_4 = \begin{pmatrix} 1 & 0 & 0 & 0 \\ 0 & 1 & 0 & 0 \\ 0 & 0 & -1 & 0 \\ 0 & 0 & 0 & -1 \end{pmatrix}$$

The simplest solution is again a simple wave, Eq. (29).

The four linearly independent general solutions of the Dirac equation are $(i = 1, 2, 3)$, $(a = 1, 2, 3, 4)$:

$$\psi_a^{(1)} = a_\alpha^{(1)} \exp\left[i(k_l x_l - \lambda x_0)\right]$$

$$\psi_a^{(2)} = a_\alpha^{(2)} \exp\left[i(k_l x_l - \lambda x_0)\right] \tag{41}$$

$$\psi_a^{(3)} = a_\alpha^{(3)} \exp\left[i(k_l x_l + \lambda x_0)\right]$$

$$\psi_a^{(4)} = a_\alpha^{(4)} \exp\left[i(k_l x_l + \lambda x_0)\right] \tag{42}$$

with the values of the coefficients $a_\alpha^{(i)}$ correspondingly defined. The general solutions can be again represented by a four-dimensional Fourier integral. The initial value problem can be solved by means of the spinor functions $S(x)$, (Ref. 4, p. 158).

Noninteracting vector, pseudo-vector, and pseudo-scalar fields. A complex vector field $\psi_\nu(x)$, $[\nu = 1, 2, 3, 4]$, is described by the action functional:

$$W = \int_{\sigma_1}^{\sigma_2} \mathcal{L}^{(1)} \, dx$$

$$\mathcal{L}^{(1)} = -\frac{1}{2} (\psi_{\nu,\mu}{}^* - \psi_{\mu,\nu}{}^*) (\psi_{\nu,\mu} - \psi_{\mu,\nu}) - \mathcal{K}^2 \psi_\nu{}^* \psi_\nu$$

(43)

$$\psi_4 = i \psi_0; \qquad \psi_4{}^* = i \psi_0{}^*$$

The corresponding Euler-Lagrange equations are

$$\psi_{\nu,\mu\mu} - \psi_{\mu,\nu\mu} - \mathcal{K}^2 \psi_\nu = 0 \qquad (44)$$

This is a set of four complex equations for the four complex components of the vector ψ_μ. Differentiating these equations with respect to x_ν furnishes:

$$\mathcal{K}^2 \psi_{\nu,\nu} = \psi_{\nu,\mu\mu\nu} - \psi_{\mu,\nu\mu\nu} = 0; \quad \text{or} \quad \psi_{\nu,\nu} = 0$$

This leads immediately to the form:

$$(\square - \mathcal{K}^2) \, \psi_\nu = 0 \qquad (45)$$

Hence the technique, explained previously, can be applied. In a similar way, other fields can be treated.

Externally interacting scalar (or pseudo-scalar) fields. Let us assume the former case of a scalar or pseudo-scalar field, but described by an inhomogeneous equation

$$(\square - \mathcal{K}^2) \, \psi(x) = \rho(x) \qquad (46)$$

with ρ refering to the interaction with external sources. This equation can be derived from a local variational principle

$$W^{(1)} = W_0 + W'; \quad W' = \int_{\sigma_1}^{\sigma_2} \mathcal{L}' \, dx; \quad \mathcal{L}' = -\{\rho\psi^* + \rho^*\psi\} \qquad (47)$$

and W_0 is the action functional of free fields given by Eq. (36). A particular solution of Eq. (46) may be furnished by means of an arbitrary particular solution of the equation

$$(\square - \mathcal{K}^2) \, G(x) = \delta(x) \qquad (48)$$

with $\delta(x)$ being a four-dimensional Dirac delta function:

$$\delta(x) = \delta(x_1)\delta(x_2)\delta(x_3)\delta(x_0) = (2\pi^4)^{-1} \int \exp{(i\,k_\mu x_\mu)}\,dk \qquad (49)$$

Eq. (46) is satisfied by the function

$$\psi(x) = \int_{\sigma_1}^{\sigma_2} G\,(x - x')\,\rho(x')\,dx' \qquad (50)$$

inside the domain Ω contained between the two boundary surfaces σ_1 and σ_2. Using the function $G(x)$ one may express an arbitrary solution $\psi(x)$ of Eq. (46) by means of Eq. (50) and the boundary values of $\psi(x)$ and its normal derivative at points on the boundary surfaces σ_1 and σ_2:

$$\psi(x') = -\left(\int_{\sigma_2} - \int_{\sigma_1} \right) G(x' - x)\,X_\mu\,\psi(x)\,d\sigma_\mu$$
$$+ \int_{\sigma_1}^{\sigma_2} G\,(x' - x)\,\rho(x)\,dx \qquad (51)$$

where $x_\mu{}'$ is a point inside Ω. Another form is:

$$\psi(x') = -\left(\int_{\sigma_{+\infty}} - \int_{\sigma_{-\infty}} \right) G\,(x' - x)\,X_\mu\,\psi(x)\,d\sigma_\mu$$
$$+ \int G\,(x' - x)\,\rho(x)\,dx \qquad (52)$$

where $\sigma_{+\infty}$, $\sigma_{-\infty}$, are spacelike surfaces situated at $(+\infty)$ and $(-\infty)$, respectively, and the volume integral extends over the whole of space–time.

Next, one may consider the interaction with external fields described by the integro-differential equation:

$$(\Box - \mathcal{K}^2)\,\psi(x) = \lambda \int_{\sigma_1}^{\sigma_2} K(x, x')\,\psi(x')\,dx' \qquad (53)$$

$(\lambda = $ coupling constant), derivable from the nonlocal action functional

$$W = W^{(1)} + W^{(2)}; \quad W^{(2)} = \int_{\sigma_1}^{\sigma_2} \int \mathcal{L}^{(2)}\,dx\,dx' \qquad (54)$$

$W^{(1)}$ being given by Eq. (36) and

$$\mathcal{L}^{(2)}(x,x') = -\frac{1}{2}\lambda\{\psi^*(x)\,K(x,x')\psi(x') + \psi^*(x')\,K(x',x)\psi(x)\} \quad (55)$$

with the kernel K satisfying the hermicity relation $K^*(x,x') = K(x',x)$, which ensures that W is real. For solutions of the perturbed equation [Equation (53)] we may use the equivalent integral equation

$$\psi(x) = \psi^0(x) + \lambda \int_{\sigma_1}^{\sigma_2} N(x,x')\,dx' \cdot \psi(x') \quad (56)$$

$$N(x,x') = \{GK\}(x,x') = \int_{\sigma_1}^{\sigma_2} G(x - x'')\,dx'' \cdot K(x'',x') \quad (57)$$

with $\psi^0(x)$ being a general solution of the interaction-free equation:

$$(\Box - \mathcal{K}^2)\,\psi^0(x) = 0 \quad (58)$$

Equation (56) is an equation of the Fredholm type and can be treated by means of the standard methods provided that the kernel $N(x,x')$ and the function ψ^0 are continuous in their arguments in the domain Ω. Another way of treating Equation (56) is by means of Picard's iteration process:

$$\psi(x) = \sum_{n=0}^{\infty} \lambda^n \psi^{(n)}(x);$$

$$\psi^{(n)}(x) = \int_{\sigma_1}^{\sigma_2} N(x,x')\,dx' \cdot \psi^{(n-1)}(x'); \quad (59)$$

$$\psi^{(0)}(x) = \psi^0(x)$$

or expressing the n th approximation in terms of the lowest:

$$\psi^{(n)}(x) = \int_{\sigma_1}^{\sigma_2} \int \ldots \int N(x, \xi^{(1)})\,d\xi^{(1)} N(\xi^{(1)}, \xi^{(2)})\,d\xi^{(2)}. \ldots$$

$$(60)$$

$$\ldots d\xi^{(n-1)} N(\xi^{(n-1)}, x')\,dx' \cdot \psi^{(0)}(x')$$

The final form is

$$\psi(x) = \psi^0(x) + \lambda \int_{\sigma_1}^{\sigma_2} R(x, x')\,dx' \cdot \psi^0(x') \quad (61)$$

$$R(x,x') = \sum_{n=1}^{\infty} \lambda^{n-1} \{N^n\}(x, x') \quad (62)$$

where $\{N^n\}$ is the matrix product of the n functions N.

Concerning the initial value problem, one may note that Eq. (61) may be used for the solution of this problem, with ψ^0 expressed by means of its initial values and the initial values of its first normal derivative, remodelled into the form:

$$\psi(x) = \int_\sigma [\Delta\,(x - x') + \lambda\{R\,\Delta\}\,(x, x')]\,X_{\mu}{}'\psi^0(x') \cdot do_{\mu}{}' \qquad (63)$$

In the above presentation, the Picard iteration process serves simultaneously for the proof of the existence and uniqueness of the solution.

Other fields. The general aspects of the technique outlined above may be applied to more complicated field equations, e.g.,

(a) externally interacting bispinor fields (generalized Dirac's equations):

$$(\gamma_{\mu}\partial_{\mu} + \mathcal{K})\,\psi(x) = \rho(x) \qquad (64)$$

$$(\gamma_{\mu}\partial_{\mu} + \mathcal{K})\,\psi(x) = \lambda\phi(x)\,\psi(x) \qquad (65)$$

$$(\gamma_{\mu}\partial_{\mu} + \mathcal{K})\,\psi(x) = \lambda \int_{\sigma_1}^{\sigma_2} K\,(x, x')\,dx' \cdot \psi(x') \qquad (66)$$

(b) mutually interacting fields (phenomena leading to equations analogous to those in electromagnetic fields interacting with an electron field);

(c) fields associated with the phenomena described by the Fokker variational principle;

(d) nonlinear interaction fields, expressible by means of the action functional:

$$W = W^{(1)} + W^{(3)} \qquad (67)$$

$$W^{(1)} = \int_{\sigma_1}^{\sigma_2} \mathcal{L}^{(1)}\,dx \qquad (68)$$

$$\mathcal{L}^{(1)} = -\frac{1}{2}\,\{\phi_{,\nu}{}^2 + m^2\phi^2\} - \{\psi_{,\nu}{}^*\psi_{,\nu} + \mathcal{K}^2\,\psi^*\psi\} \qquad (69)$$

$$W^{(3)} = \iint_{\sigma_1}^{\sigma_2} \int \mathcal{L}^{(3)}(x', x'', x''')\,dx'dx''dx''' \qquad (70)$$

$$\mathcal{L}^{(3)} = -\frac{\lambda}{3!} \sum_{\text{perm}} M(x', x'', x''',) \, \psi^*(x') \phi(x'') \psi(x''') \tag{71}$$

the sum $\sum\limits_{\text{perm}}$ being the sum over all permutations of the three points x', x'', x''' in the integrand of Eq. (71). The corresponding Lagrangian (field) equations are

$$(\square - m^2)\phi(x) = \lambda \int_{\sigma_1}^{\sigma_2} \int M(x', x, x'') \psi^*(x') \psi(x'') \, dx' dx'' \tag{72}$$

$$(\square - m^2)\psi(x) = \lambda \int_{\sigma_1}^{\sigma_2} \int M(x, x', x'') \phi(x') \psi(x'') \, dx' dx'' \tag{73}$$

It seems that the techniques, methods and particular cases presented above will cover a considerable number of problems in epihydrodynamics. In general cases, the Picard iteration process may be used almost always, at least in a local sense.

Remarks on Quantum Hydrodynamics

The present remarks were put down after the symposium on physical and biological phenomena under zero g conditions. In his presentation Dr. Benedikt called the attention to a great number of forces and phenomena which never before appeared in the investigations of the classical hydrodynamic nature, but which should be taken into account in epihydrodynamics. To mention only some of them, they are of the following character:

Rest Conditions

Valence
Van der Waals forces
Average potential energy of intermolecular forces
Thermodynamic definition of surface tension
Kinetic pressure = flux density of molecular momentum
  cohesion = flux density of intermolecular forces
Temperature effect upon surface tension and energy;

Motion Conditions

Boussinesq surface viscosity   Eyring's viscosity in liquids
Inception of capillary phenomena   Electrocapillary effects
Inception of conductive heat transfer   Barocapillary effects, etc.

These phenomena may be subject to very rapid time variations near the zero g conditions; therefore, the question of the precision and the accuracy in the mathematical description of these phenomena is becoming important. Let us give a brief review of the possible philosophies and methods available today for the mathematical description of such phenomena in hydrodynamics:

(i) Phenomenological descriptions, based usually upon the direct application of Newton's law to the system in question. Here one has the Euler equation and the Navier-Stokes formulation; one may maintain that it should be possible to include all the phenomena, mentioned above, into the macroscopic formulation of these equations by means of some macroscopic coefficients, properly selected and adjusted to the conditions in question; as it seems acceptable that such a description may be possible, it is an open question whether this is the best possible description. Moreover, one must keep in mind that it will be necessary to collect and investigate a great number of the coefficients, which, being fundamentally of the microscopic (quantum) character, must be transformed into a macroscopic form to be inserted into the macroscopic equations of motion.

(ii) The classical mechanics description of the hydrodynamic system in question. This problem is discussed thoroughly in the previous sections of the present paper. The classical field theory belongs here as well; the classical description was chosen by Benedikt as his tool.

(iii) Statistical classical mechanics (with Liouville theorem) referring to systems of f (finite) degrees of freedom.

(iv) The classical kinetic theory of gases in the Maxwell-Boltzmann formulation.

(v) The quantum statistical mechanics (with the analogue of Liouville's theorem).

(vi) The statistical fluid dynamics discussed above is not yet in a suitable form for attacking the problems in fluid dynamics.

(vii) The quantum hydrodynamics. Below we shall discuss briefly some aspects of this philosophy.

The field of quantum hydrodynamics can be divided into three subfields, i.e., (i) the concept of the continuum; (ii) the concept of discrete particles; (iii) special methods (e.g., Landau's theory of Helium II, etc.). According to the opinions of many theoretical physicists, the most promising seems to be concept (ii), the least promising, concept (i). The general philosophy of the quantum hydrodynamics is in some sort of controversial status. One opinion is that this philosophy can be applied only to low-temperature phenomena (Helium II); on the other hand, Pines applied the concepts of quantum hydrodynamics to the plasma dynamics. It is too early today to speculate whether the concepts of quantum hydrodynamics may be applied to epihydrodynamics, and if so, in what form. It should be emphasized that one witnesses today an increasing rate of the application of quantum hydrodynamics concepts to some specific aspects of hydrodynamics like transition phenomenon, etc. The concept of long- and short-range forces, discussed by Benedikt, shows clearly that the classical Boltzmann equation (with a single sphere of a strong interaction and of short-range forces) cannot be applied. Some time ago the author of the present paper and one of his coworkers[5] constructed a philosophy of a multiple-sphere interaction (short- and long-range (Coulomb) forces) in quantum gas-dynamics, obtaining a system of intercorrelated and interdependent Boltzmann-like equations. A great number of new phenomena, forces and coefficients of a microscopic, quantum nature (even the intermolecular forces, if treated properly, should be formulated from the quantum point of view) introduced by Benedikt give rise to a doubt whether the methods of classical mechanics (including the classical field theory, the classical kinetic theory of gases) are strong enough to describe mathematically the various phases of epihydrodynamics. On the other hand, the quantum hydrodynamics formulation is in a very vague status and a great amount of energy has to be devoted to constructing the proper fundamentals before some workable schemes can be proposed. It is absolutely premature to speculate today whether the quantum hydrodynamics will or will not be an applicable tool to epihydrodynamics; a certain amount of the fundamental investigations must be done before some definite conclusions can be derived.

Concluding Remarks

The author presents two kinds of philosophy on the possible treatment of equations of epihydrodynamics, derived by Benedikt; first, the classical field theory technique, second the statistical fluid dynamics proposition. As the first technique is developed and can be applied, the second is still in the status of constructing the fundamentals. Using the first technique, the author discusses some types of the action functionals and of the field equations which may possibly occur in epihydrodynamics formulation. A remark on the general Picard's iteration procedure applicable almost in all the cases and a discussion on the possibility of the application of the quantum hydrodynamics tool to epihydrodynamics close the paper.

References

1. E. T. Benedikt, "Scale of Separation Phenomena in Liquids Under Conditions of Nearly Free Fall," *Jour. Amer. Rocket Soc.*, 29, 1959, pp. 150-151.

2. E. T. Benedikt, "Epihydrodynamics: The Dynamics of Liquids Under the Action of Surface Tension Forces" presented at The Divisional Meeting of the Division of Fluid Dynamics, Amer. Phys. Soc., University of Michigan, Ann Arbor, Mich., 1959.

3. J. Rzewuski, "Field Theory, Part I, Classical Theory," Polish Monographs, Warsaw, Panstwowe Wydawnictwo Naukowe, 1958.

4. S. S. Schweber, H. A. Bethe, and F. de Hoffmann, *Mesons and Fields*, Volume I, Fields Row, Peterson and Com., Evanston, Ill., 1955.

5. C. Whittenbury and M. Z. v. Krzywoblocki, "On Certain Fundamental Concepts of Classical and Quantum Kinetic Theory of Cases," *Acta Physica Austriaca*, Vol. XIII, 4, 1960, pp. 395-472.

BEHAVIOR OF LIQUIDS IN FREE FALL*

WILLIAM C. REYNOLDS†

*Assistant Professor of Mechanical Engineering,
Stanford University, Stanford, California*

Recent rocket and satellite activities have brought into focus a number of important problems involving the behavior of liquids under weightless conditions. Study of the behavior of liquids in free fall can give considerable insight into the behavior of liquids in space. A photographic study of boiling during 8 ft. of free fall[1] clearly demonstrated the existence of differences between earthbound and and weightless fluid behavior. An analysis of the relative orders of magnitudes of surface tensile, viscous, and inertia forces acting on weightless liquid[2] allows estimation of the approximate time required for deformation due to surface tension after removal of gravity forces. In the present note, some photographs of fluids in free fall are included. It should be noted that this study was entirely qualitative and of a very preliminary nature. For these experiments, a free-fall vehicle consisting of a 2-in. square lucite box and a heavy steel weight was constructed. The vehicle was dropped from heights up to 16 ft. and caught by an arresting system. Liquids placed in the lucite box were photographed at the bottom of their fall by a 3-microsecond exposure flash camera. While the container was dropped from successively greater heights, a series of photographs was taken which show the progress of the fluid in free fall. It is important to remember that the photographs were taken during different drops and, therefore, are not quite what would appear in a series of motion-picture frames. The deceleration due to wind resistance was estimated at a maximum of 0.01g, and, therefore, the fluids were almost exactly weightless.

In the first series of drops, the floor of the lucite box was lined with paraffin, and a small amount of mercury placed on the paraffin.

*Reprinted from *Journal of the Aero/Space Sciences*, Vol 26, No. 12 (Dec. 1959) by courtesy of the Institute of the Aeronautical Sciences.

†In the preparation of this note, the assistance of Stanford students, G. A. McKinzie, J. Gerstley, H. Munson, A. Lindsay, N. Wooldridge, O. Kyte, and A. Hill is gratefully acknowledged.

The box and the mercury, before the drop, can be seen in the upper left corner of Figure 1. Note that the mercury is off to the left side of the box. Photographs taken after increased durations of free fall are shown. Note that the surface forces act very quickly to pull the mercury together, once the gravity forces are removed. In fact, the surface energy is sufficient to lift the mercury glob completely off the surface, where it hangs in space, as shown. For a short time the glob retains an irregular shape, but before long the surface forces start to reshape the glob into a sphere. The time estimates for deformation are consistent with those predicted by

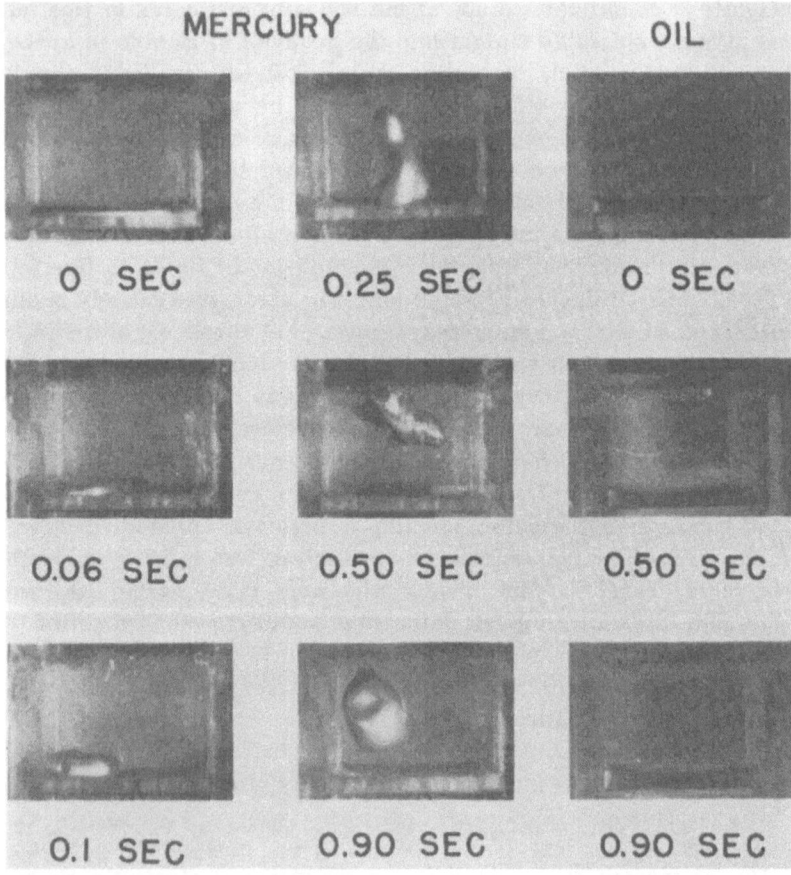

Figure 1. Liquids in rectangular container under conditions of free fall.

Benedikt.[2]   A similar series of tests with mercury on the lucite floor yielded essentially the same results.

Photographs taken during a series of drops with oil on the lucite floor are shown in Fig. 1.   Because of the relatively low surface tension for the oil, the time required for motion is somewhat longer than for the mercury; but even in the short time available, some motion occurs.   Note that the oil does not tend to globulate, but instead starts immediately to climb up the wall.

The difference between the behavior of mercury and oil can be attributed to the type of wetting.   Mercury does not wet the floor, and therefore surface forces tend to pull it together.   Oil does wet the wall, and the surface forces tend to pull it up the wall.

On the basis of these experiments, it is concluded that fluids which wet their container will crawl around the wall, leaving the gas pocket in the center.   Fluids which do not wet will kick themselves off the wall, leaving the gas in contact with the wall. However, from energy considerations, it is evident that the wetting fluids cannot crawl forever; the surface area is limited by the energy available for stretching it, and therefore a wetting fluid may not completely cover its container.   Similarly a non-wetting fluid may tend to form more than one glob; the energy available for moving the mass off the floor may not raise the centroid enough to allow a sphere to form, in which case the ultimate shape would probably be two or more spheres.

References

1. R. Siegel and C. Usiskin, "A Photographic Study of Boiling in the Absence of Gravity," *J. Heat Transfer*, Vol. 81, Series C, No. 3, August 1959, p. 230.

2. E. T. Benedikt, "Scale of Separation Phenomena in Liquids Under Conditions of Nearly Free Fall," *J. ARS*, February, 1959. See also article "General Behavior of a Liquid in a Zero or Near-Zero Gravity Environment" in this volume.

EXPERIMENTAL PRODUCTION OF A ZERO OR NEAR-ZERO GRAVITY ENVIRONMENT

E. T. BENEDIKT and R. LEPPER

Head, and Senior Engineer, Space Physics Laboratory, Astro Sciences Group, Norair Division, Northrop Corporation, Hawthorne, California

In view of the peculiar nature of gravitational and inertial forces, a zero gravity environment can be approximated under terrestrial conditions only within falling systems. As a consequence of theoretical results established in Refs. 1, 2, it follows that in order to produce observable hydrodynamic phenomena typical of low or zero gravity conditions within an enclosure falling in the atmosphere, the load factor (that is, the aerodynamic deceleration per unit weight) of the descending system must be less than a critical value

$$n_* \simeq \sigma/\rho g L^2$$

where σ, ρ are respectively the coefficient of surface tension and the density of the liquid, $L \simeq \sqrt[3]{3V/4\pi}$ the linear dimensions of the volume, V, occupied by it, and g, the acceleration of gravity. Numerical values of $n_* L^2$ for various liquids of interest appear in Table 2, Ref. 2. Actual numerical examples, illustrating the effect of size upon n_* for water and mercury are given in Table I of the present paper. From these numerical data it appears that the critical load factor is of the order of 0.01. Consequently, load factors considerably smaller than this value, must be achieved for purposes of laboratory experimentation. Insomuch as the load factor of a system falling within the atmosphere builds up very quickly to its equilibrium value of unity,* the time through which an adequately low gravity environment can be obtained within the system

---

*By means of elementary considerations, it is easy to show that the kinematics of a body falling vertically under the retarding action of an atmosphere, approximates very closely that of a free vertical fall as long as the distance z through which the body has descended (starting from rest) is much less than a critical length $\lambda = m/C_D \rho_0 A$, m being the mass of the body, A a reference cross section in a direction normal to the motion, C_D a suitable average value of the drag coefficient, and ρ_0 the density of the atmosphere. Under these conditions, the ratio z/λ provides a very good estimate of the load factor.

is limited. The limit depends upon the smallness of the load factor desired and upon the size and mass of the system; however, it can be stated that in all practical situations, this limit is certainly less than one second. It must also be kept in mind that the duration of the experiment must exceed the period of time.

$$t_* \simeq \sqrt{\rho L^3 / \sigma}$$

required by the liquid to effect the transition from the configuration which it assumes in the terrestrial gravitational field, to that pertaining to the absence of any gravitational or inertial field. [1,2]

As shown in Refs. 1, 2, numerical values of $t_*/L^{3/2}$ for various liquids appears in the already mentioned Table 2 of Ref. 2. Numerical examples for samples of water and mercury of typical size are shown in Table I. It is thus apparent that it is necessary to impose a lower limit to the height of fall of the experimental system regardless of whether the system is dropped in a vacuum or in the atmosphere. In practical cases, t is of the order of 0.2 sec corresponding to a height of fall of about 0.6 ft. Of course, additional time (i.e., a greater length of fall) has to be provided to eliminate the after effect of the above transition.

TABLE 1: DYNAMIC BEHAVIOR OF WATER AND MERCURY IN LOW GRAVITY ENVIRONMENT

Water: Specific Surface Tension = 72.85 cm^3/sec		
	$L = 5cm$	$L = 1m$
Critical acceleration	.00297$_g$	.00000743$_g$
Duration of transient	1.31sec	1.95min
Mercury: Specific Surface Tension = 37.73 cm^3/sec		
	5cm	1m
Critical acceleration	.00154$_g$	.00000385$_g$
Duration of transient	1.82sec	2.72min

Finally, it should be stressed that the conditions obtaining within the liquid sample at the instant of release of the cell must be determined if interpretable data are to be obtained. In general, a condition of rest—i.e., equilibrium of the surface as well as absence of any circulatory motion inside the liquid—is the one most easily determined. Such a condition can be attained in practice by allowing the liquid sample to remain undisturbed in its starting position for a length of time sufficient to permit decay (through viscous dissipation of energy) of all initially present circulatory currents. For samples of centimeter dimensions, a period of rest of between 30 min to 1 hr appears adequate.

The Method of the Encapsulated Cell

Application of the basic requirements discussed above indicates that adequately low gravity environments can be produced within systems dropped in the atmosphere only for comparatively short times. Some successfully operating devices of this nature, and the results obtained with them are described elsewhere in this volume. However, devices of rather awkward weights and dimensions would be necessary to obtain, by this method, a very close approach to rigorous zero gravity conditions for more prolonged durations. Such conditions can thus be attained most simply within enclosures falling in evacuated surroundings. The most direct way of achieving this would consist of the use of an evacuated vertical tube through which a test cell could fall from and into suitable airlocks. On the basis of the principles reviewed in the preceding article, a free fall having a duration of between 2 and 3 sec can be regarded as sufficiently long to permit full development and observation of zero gravity hydrodynamic phenomena on a scale of the order of 1 cm. The required height of the tube would thus be of the order of 100 ft. The operation of such a device would involve a lengthy period of evacuation and complicated manipulation of the airlocks and other difficulties.

These shortcomings are avoided by the method of the encapsulated cell (already briefly described in Ref. 3). The principle of this method (see Figure 1) consists of enclosing the test system –

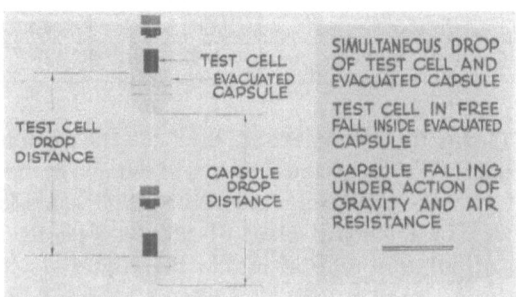

Figure 1. Principle of encapsulated cell.

that is, essentially a cell containing the liquid to be observed with some necessary instrumentation attached—in an evacuated capsule which can be dropped within the conventional atmospheric environment. The actual configuration of an instrumented package embodying the principle of the encapsulated cell appears in Figure 2. If the inner test system is simultaneously released, it will be falling under conditions simulating extremely closely those of free fall.

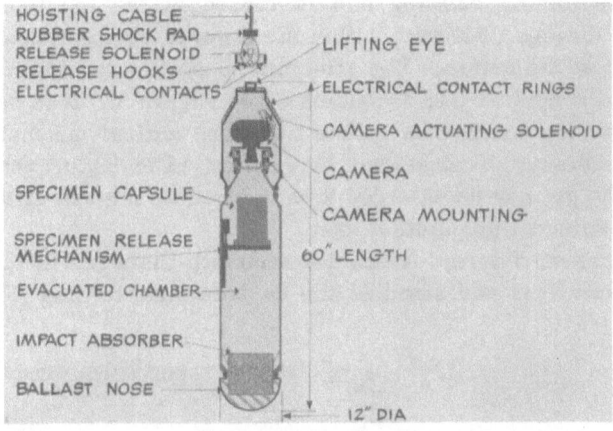

Figure 2. Cross section of Norair zero gravity capsule.

Actually, in view of the low velocity of the inner cell relative to the outer capsule, load factors sufficiently low to bring about zero gravity phenomena can be obtained even by dispensing altogether with evacuation of the outer capsule. In any event, the freely falling test system would gradually overtake and collide with the bottom of the capsule; in practice, for drops from heights not exceeding some hundred feet, and with capsules of reasonable length, this event would be preceded by the termination of the experiment.*

*This is due to the fact that the motion of the cell relative to the capsule is quite slow in the initial phases of the drop. It can be shown that as long as the distance and way which the outer capsule has desended is much less than λ (see previous footnote), the displacement of cell relative to the (evacuated) capsule is of the order of z^2/λ.

Among other advantages the above arrangement permits to isolate the test cell from recording instruments such as cameras, which can be mounted in the capsule. The undesirable effects of inertial forces and torques which would be imparted to the liquid if instruments containing moving parts were rigidly connected with the free test cell are thus avoided.

Norair Zero Gravity Tower

Structure. The tower (Figures 3, 4) consists of a vertical cylinder 87 ft long over-all and 24 in. in diameter. The structure consists of 81 ft of steel above ground (the height of a six-story building) and a pit 6 ft below ground level. The pit is used for deceleration provisions. The steel portion is designed to stand unsupported; however, it is stabilized by being tied into the structure of the adjoining building and is closed at the top to minimize weather damage. All work within the tower is accomplished through the door at the bottom. The structure supporting the tower required a door of somewhat odd dimensions (see Figure 5). It is semicircular at the top and bottom with a maximum vertical dimension of 44 in.; the horizontal dimensions tapers from 18 to 12 in. On a future model the pit may be enlarged to accommodate ingress at that point from an adjacent preparation room.

Alignment fixture. A special structure installed in the top of the tower aligns and steadies the capsule prior to drop (Figure 6).

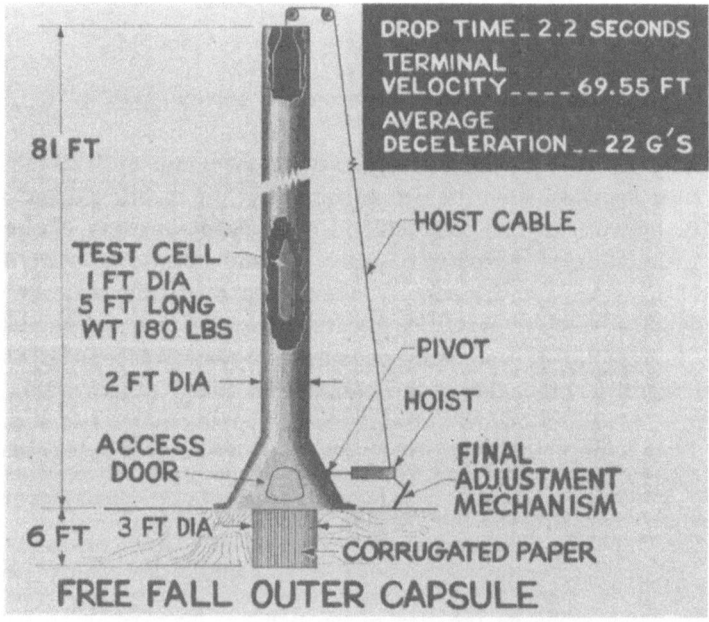

Figure 3. Schematic of Norair zero gravity tower.

Figure 4. Norair zero gravity tower installation.

Figure 5. Base of Norair zero gravity tower.

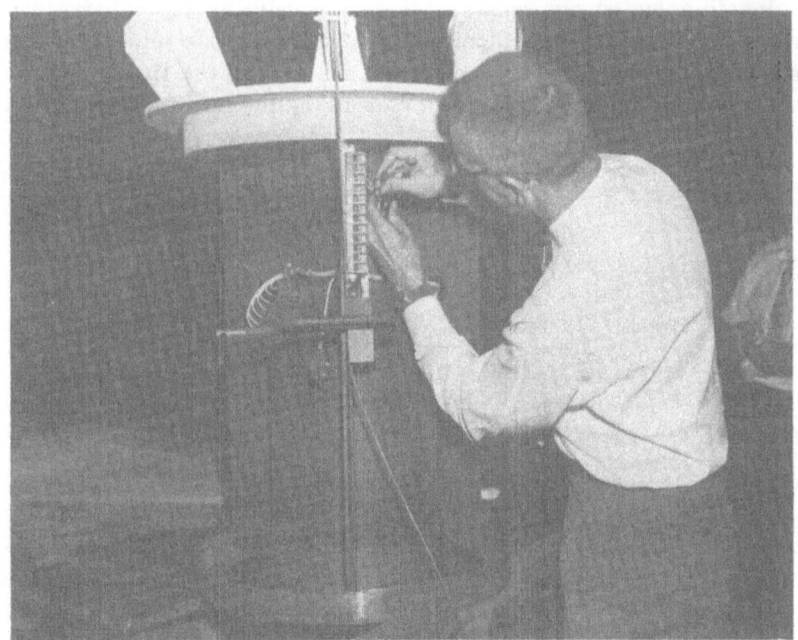

Figure 6. Capsule alignment fixture prior to installation in top of zero gravity tower.

The capsule is guided for the first inch of travel after release to further insure that no angular velocity is imparted to the capsule during the release. This structure also contains the electrical connections to the capsule and the release mechanism.

Hoist mechanism. After insertion at the bottom of the tower, the capsule is hoisted to the top by electric motor and cable. The raising speed is set very low (24 ft per min) to minimize the chance of disturbing the experiment prior to the actual drop. Safety circuits within the hoist itself prevent any dropping of the load in the event of electrical power loss. A limit switch stops the hoist approximately four in. from the top. Final adjustment and seating of the capsule in the alignment fixture is accomplished by a screw jack which raises or lowers the hoist motor. Proper seating of the capsule is indicated by a light at the hoist control panel. This light also demonstrates that continuity is established in the circuits within the capsule, and release may be accomplished.

Hoist cable release. The quick-release mechanism (Figure 7) is a standard piece of Norair test equipment of proven design. It is solenoid-actuated and opens in such a manner as to minimize any reaction in the dropped specimen. By simultaneous release of two stops, the hook is sprung open from each side of the lifting eye, no side force is exerted against the capsule by the release mechanism.

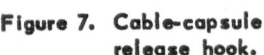

Figure 7. Cable-capsule release hook.

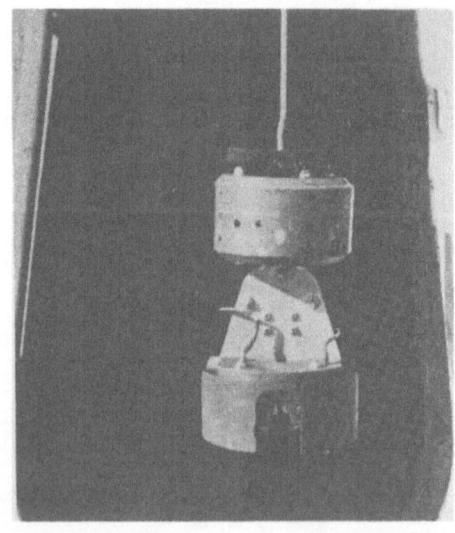

The extremely short duty cycle allows the solenoid to be operated at considerably above rated voltage thus insuring immediate, positive action.

Deceleration material. The material presently being utilized to limit the deceleration forces at termination consists of spirally wound corrugated paper set on edge (Figure 8). This then acts as a number of parallel thin-walled tubes and results in a relatively constant deceleration force. The tightness with which it is packed into the pit determines its resistance to buckling, thus controlling the force required for crushing. As arranged in the present installation, the equipment is being subjected to less than 25 g's during deceleration. For a future installation other nondestructive means, such as pneumatic or hydraulic systems, are being considered.

Figure 8.   Dummy capsule embedded in deceleration material. Capsule, containing an accelerometer, was used to verify impact deceleration prior to test with actual instrumented capsule.

Instrumented Package

General assembly. The capsule (Figures 2, 9 and 10) is 60 in. long over-all, 12 in. in diameter, and weighs 180 lb. It is aerodynamically shaped for stability during the drop. The nose section is a partially hollowed steel hemisphere. Steel was used to withstand the shock at the termination of the drop and as ballast to maintain the center of gravity 15 in. from the nose (25% point). Further adjustment of the center of gravity is possible by drilling the

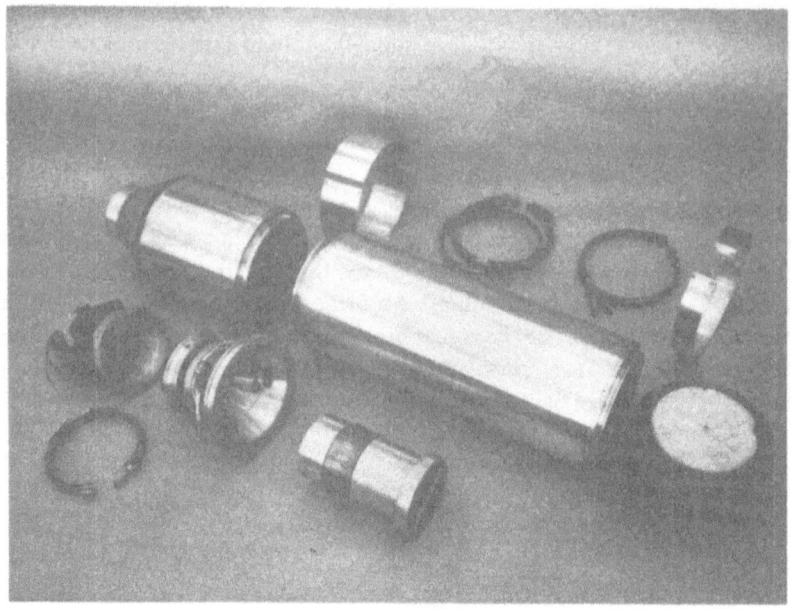

Figure 9. Components of Norair zero gravity capsule.

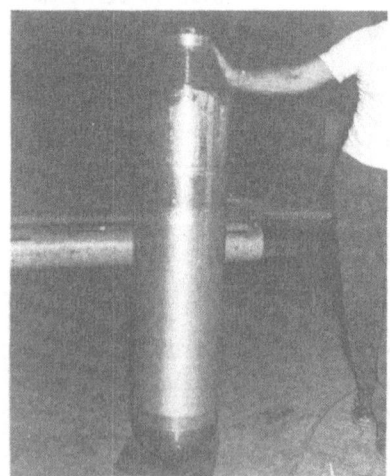

Figure 10. Norair zero gravity capsule assembled.

Figure 11. Zero gravity capsule camera mount.

nose section from within the capsule and removing material, or by filling such holes with lead.

The skin of the outer capsule center section is of aluminum, of a thickness capable of withstanding the pressures due to evacuation of the capsule and the shock of termination of the drop. The length of this section determines the duration of free drop time of which the assembly is capable.

The top portion of the chamber, generally evacuated down to one micron of mercury, consists of a plate glass lens and camera support (Figure 11). It is through the plate glass lens that the camera views the test cell and records the reactions of its contents. The camera mount is steadied throughout the descent by spring-loaded balls in detents. Upon contact of the capsule with the deceleration material, the balls raise out of the detents, transferring the major portion of the shock to be absorbed to the rubber mounting material. The aft portion or hood of the capsule carries all of the electrical contacts for the release sequence. Power and return leads are carried up the outside of the tower to contactors in the alignment fixture. On proper seating of the capsule, electrical contact is made through the contactors to copper rings inserted into the hood.

Camera. It was determined that the lens criteria should include a circle of confusion not greater than 1/200 in. for any distance between 6 and 30 in. from the camera (the total travel of the test cell), using but one lens setting. An 18.5-mm lens in a 35-mm movie camera satisfied these requirements, and is utilized in this installation. This camera is spring-driven to eliminate a battery requirement in the outer capsule.

Test cell. The specimen is located in a small clear cup within the test cell (Figure 12). The top of this cell is also a lens through which the camera shoots. Four mirrors are located so as to allow five views of the specimen to be seen simultaneously; one direct view and one from each side through the mirrors (Figure 13). The pictures so obtained appear as shown in Figures 15 and 16. The fabrication of the clear cup required the use of light dams in the areas above the light inputs. To further utilize all light available, the dams and bottom of the cup are silvered. The inside dimensions of the cup, the actual specimen area, are one inch in diameter and 3/4 in. high. It is illuminated from below by four 60-watt bulbs, the light being carried to the specimen through Lucite lenses. The bulbs are powered by batteries located in the section

Figure 12. Zero gravity test cell, disassembled.

Figure 13. Zero gravity test cell.

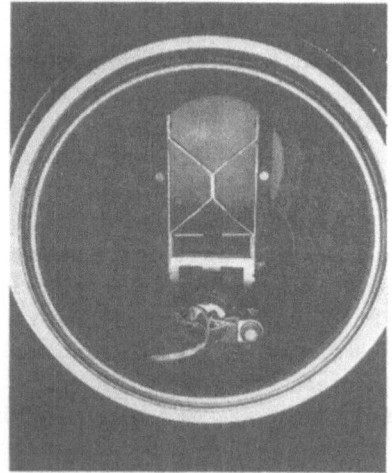

Figure 14. Test cell release mechanism, cocked in support-
ing position.

immediately below the bulbs of the test cell. They are lighted by means of a magnetically self-holding relay actuated through contactors in the base of the test cell. A gravity switch extinguishes the bulbs at the termination of the test. This is both to conserve the limited life of the batteries and bulbs and to prevent excessive heating of the lenses and equipment in the test cell.

Cell release mechanism. The test-cell release mechanism (Figure 14) is located on the wall of the outer capsule. This device holds the test cell in the top of the evacuated chamber until the start of the drop. At this point the support is removed from under the test cell in such a manner as to impart no velocities to the container. It is solenoid-operated and spring-loaded. The support moves straight down approximately 3/4 in. and then folds and locks against the side of the outer capsule, out of the way of the test cell as it drops. The surface on which the cell rests prior to drop is shock- and vibration-isolation material to further safeguard the test specimen from external forces. Imbedded in this material are the contactors for actuating the lamp control relay in the test cell. The residual atmosphere (corresponding, as already noted, to a pressure of one micron of mercury) produces a load factor not exceeding about 10^{-6}. During the drop the displacement of the cell relative to the capsule has been estimated (on the basis of an approximation indicated in a previous footnote) to be of the order of a few inches.

Cell deceleration material. Since some tolerance must be allowed in the timing of the various mechanical release mechanisms, the inner dimension of the capsule was designed larger than required. This means that the inner test cell will not reach the bottom of the capsule during the drop when their difference in velocities is small. Rather, the capsule will be decelerating rapidly when contact is made. Therefore the nose section of the capsule contains shock absorption material to cushion this contact.

Operation. The release sequence is such that the test cell lamps are energized and the camera started immediately on actuation of the drop switch. Approximately one second later the test cell release mechanism is actuated. This delay is to allow the lamps to fully heat up prior to start of test and to obtain "zero time" pictures. The delay is governed by a thermal time-delay relay in the control panel. Approximately one 1/100 sec after the test cell release mechanism is actuated the outer capsule is released. This delay is obtained by inserting an additional relay in the capsule release circuitry. Since the capsule drops away from the contactor in the alignment fixture, only self-contained equipment can be electrically started using these connections. Such equipment may be mechanically driven, or batteries may be installed in either the capsule or test cell or both.

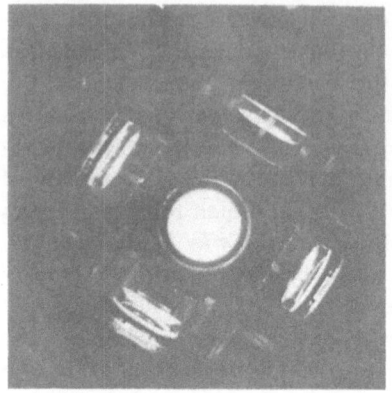

A. Time Zero (Water Under Action of Terrestrial Gravity)   **B. 0.032 Sec After Start**

C. 0.288 Sec After Start   **D. 0.416 Sec After Start**

Figure 15. Portions of test sequence showing typical configurations of water corresponding to transition from normal to zero gravity situation.

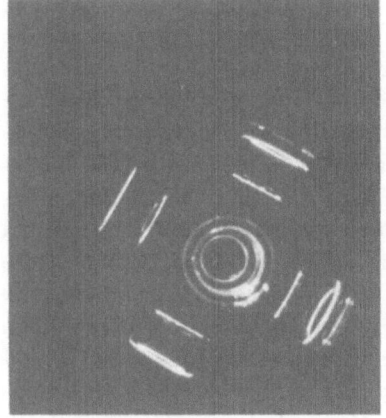

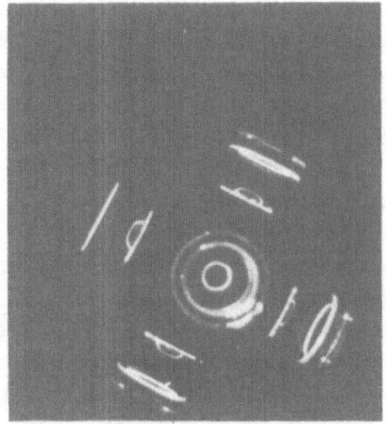

A. Time Zero (Mercury Under Action of Terrestrial Gravity)

B. 0.144 Sec After Start

C. 0.216 Sec After Start

D. 0.612 Sec After Start

Figure 16. Portions of test sequence showing typical configurations of mercury corresponding to transition from normal to zero gravity situation.

Typical Results

Pictures obtained in test using water and mercury are shown in Figures 15 and 16 respectively. They show various stages of the configuration of the fluid on being subjected to an abrupt transition

from normal to zero gravity environment. Figure 15a was taken at
the instant of release as indicated by light just coming on at the
upper right-hand border of the picture; note the horizontal portion
of the meniscus. Figure 15b, taken 0.032 sec after the start of the
drop, shows the meniscus beginning to depress into a concave
shape. Figures 15c and d were taken at 0.288 and 0.416 sec respectively. It should be noted in Fig. 15d that the meniscus exhibits the concave spherical shape typical (inside cylindrical
containers) of a zero gravity environment. Figure 16 shows the
results obtained with a mass of 20 grams of mercury occupying the
test cell to a depth of approximately 0.1 in. (2.6 mm). Figure 16a
shows the initial situation 0.036 sec prior to start; Figure 16b, 0.144
sec after release. The increasing curved convex meniscus is apparent in these pictures. Figures 16c and d were taken at 0.216
and 0.612 sec respectively. In the latter picture the mercury
appears as a freely floating sphere. The quantitative results of
these experiments[4] corroborate the theoretical predictions of Refs.
1, 2 and 3.

Possible Improvement of the Encapsulated Cell Method

An interesting modification* of the preceding procedures could
be obtained by overcoming the aerodynamic drag acting upon the
outer capsule with the force exerted by a cable (see Figure 17).
This cable is driven by a motor whose speed is regulated in such
a manner as to maintain the test cell always in the identical position relative to the capsule for the duration of the experiments.
The control signals would be provided by the displacements of the
test cell assembly, the latter functioning as a sensing element,
i.e., an accelerometer. The motion of the capsule would thus be
identical to that of a free vertical fall.

Such an arrangement would present a number of advantages.
Optical observation of improved quality would be afforded as a
consequence of the almost constant position which the test cell
would maintain relative to the optical system. In addition, electrical connections could be established between the ground and the
capsule without interfering with the motion of the latter, thus greater
amounts of power would become available for instrumentation; in

---

*Conceived jointly by R. Lepper and E. A. Smith of Astro Sciences
Group, Norair Division, Northrop Corporation.

particular, closed-circuit television could be installed. Also the cable system might be utilized to assist in braking the vehicle, thus relieving the problem of deceleration. This method has not yet been reduced to practice, however provisions were incorporated in the construction of the Norair zero gravity tower to allow conversion to this system if desired. In view of the so far satisfactory performance of the present system, no plans for such a conversion are presently entertained.

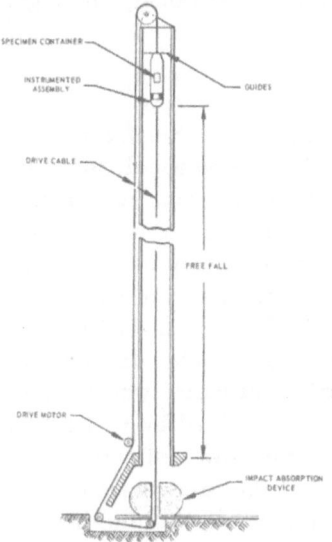

Figure 17. Zero gravity installation with servo-controlled cable-driven capsule.

Remarks Regarding the Production of Near Zero Gravity Environment in Airborne and Space Vehicles

The attainment of near-zero gravity environments in aeroplanes piloted along parabolic free-fall trajectories and results obtained by this method are discussed elsewhere in this volume.

Prolonged (20-40 sec.) periods of low gravity can thus be obtained, during which experimentation with various hydraulic and thermal devices, as well as astrobiological and psychological research have been successfully conducted. However, it is felt that conditions leading to extremely low (and steady) gravity environments required for fundamental physical zero gravity research cannot be reached by means of pilot-controlled aircraft. In addition, the liquids under examination could not be maintained in the prolonged state of rest required for the attainment of well-defined initial conditions. The latter difficulty would presumably be even more pronounced in missile-borne tests. It appears thus that only an orbiting laboratory would provide a strictly zero gravity environment (except for the effects of interplanetary matter) for an indefinite period of time. Until such a project should become a reality it appears that terrestrial production of a zero gravity environment along the lines discussed in the present report, or equivalent methods seem to offer the closest approach to a rigorously zero gravity environment.

Acknowledgements

Substantial contributions were made to the construction of the Norair zero gravity tower by various Norair personnel, to whom the authors wish to extend due credit: D. Olmore was responsible for the coordination of the project; G. Scott attended to the procurement of materials; H. Magill, H. Sarkesian and D. Troup assisted in various design problems. Stress analysis problems were handled by R. Hayes and J. Sing; R. Brunpton, L. E. Picker, G. F. Younkin, T. Williams attended to the fabrication of the capsule, its instrumentation and other components; the electrical circuitry was attended to by V. M. Urban. The installation was supervised by W. Gaulka. In addition, it should be mentioned that the tower structure, the installation of its base and the erection of the tower were carried out by the Rheem Manufacturing Co. of Downey, California, under subcontract with Norair.

References

1. E. T. Benedikt, "Scale of Separation Phenomena in Liquids Under Conditions of Nearly Free Fall," *ARS Jour.*, 29, 1959, p. 150.

2. E. T. Benedikt, "General Behavior of a Liquid in a Zero or Near Zero Gravity Environment," presented at the Norair/AAS Symposium on Zero Gravity, July 1960.

3. E.T. Benedikt, "Epihydrodynamics; The Dynamics of Liquids Under the Action of Surface Tension Forces," presented at the Divisional meeting of the Division of Fluid Dynamics, Amer. Phys. Soc., University of Michigan, Ann Arbor, Mich., 1959. Partially reproduced as ASRL-TM-59-16-Z1, Norair Division, Northrop Corporation.

4. R. Lepper, "Experimental Studies of the Hydrodynamic Behavior of Liquids in a Zero Gravity Environment," ASG-TM-61-13-Z5.

PART TWO

*Heat Transfer and Boiling Phenomena
Under Zero Gravity Conditions*

AN EXPERIMENTAL STUDY OF BOILING IN REDUCED AND ZERO GRAVITY FIELDS*

C. M. USISKIN AND R. SIEGEL

Lewis Research Center, National Aeronautics and Space Administration, Cleveland, Ohio

One of the most important environmental changes which man encounters as he leaves the earth is the change in the gravity field. In free space or in an orbiting satellite, the gravitational field may closely approach zero. On the surface of the moon or the inner planets of our solar system, the gravitational field is less than that on earth. Heat-transfer processes such as free convection, condensation and boiling depend on the gravitational body force, and hence will be affected by this new environment. As yet, to the authors' knowledge, there has been no experimental heat-transfer information available for reduced gravity fields. Hence any design calculations for this range must be based upon extrapolated theories which contain a gravity parameter, but which have only been checked experimentally at normal gravity. For very low gravity fields, these theories may become inaccurate, since forces such as surface tension, which are sometimes neglected, become increasingly important as the body force is reduced. In a previous photographic study of boiling in the absence of gravity,[1] the authors demonstrated the importance of gravity in the pool-boiling mechanism. Another photographic study on the behavior of liquids in free fall[2] has showed the significant influence of surface tension forces when the gravity field is removed.

The goal of the present work is to provide quantitative information on how reducing the gravity field affects boiling heat transfer and bubble dynamics. The latter is of some interest in connection with void formation in water-moderated nuclear reactors. Two recent semitheoretical correlations which have been proposed for nucleate boiling are those of Rohsenow and co-workers and of

---

*This paper has originally appeared as Publication No. 60 HT-10 of the American Society of Mechanical Engineers. Reproduced here by courtesy of ASME.

Forster and Zuber.[3] The correlation of Rohsenow takes the velocity of a bubble at the instant of break off from the surface as being the most meaningful velocity in the heat-transfer mechanism. Since this velocity depends on the buoyancy force of the bubble, it is not surprising that the heat-transfer coefficient is a function of gravity. However, in the theoretical correlation, gravity appears to only a small power. On the other hand, Forster and Zuber feel that the radial growth velocity of the bubbles while they are still close to the surface is more significant in governing the heat-transfer process. As a result, their heat-transfer correlation is independent of the gravity field. Hence the theories are not in agreement as to the role of gravity in nucleate boiling.

The theories for nucleate boiling do not predict the critical heat flux. An analysis of Zuber[4] on transition boiling, gives a prediction of the critical heat flux. The theory is based on hydrodynamic instability between the liquid and vapor motions near the heated surface. These motions are gravity dependent, and hence gravity appears in the final expressions. A few other theories which are summarized by Cole[5] also predict that the critical heat flux will be a function of gravity. This dependence is investigated experimentally in the present work.

The apparatus used in the present investigation was a modification of that employed in Ref. 1. The boiling equipment was placed on a platform which could be hoisted and then dropped a distance of nine feet. A counterweight was attached to the platform to slow its downward acceleration and thereby provide gravity fields between zero and one. A high-speed motion-picture camera was mounted on the platform so that boiling could be photographed for various gravity fields and heat fluxes. The motion pictures in the nucleate boiling regime were analyzed to determine the effect of gravity on bubble sizes and rise rates. Color pictures were taken of film boiling from a red-hot surface for a few different low-gravity fields. The equipment was instrumented so that a study could be made of the variation of burnout heat flux with gravity field. The burnout data for each experimental run were recorded while the platform was falling through the air. Since the fall lasted only about 1 sec, a proper interpretation of the data required a knowledge of the transients involved in the burnout phenomenon. Hence a brief investigation of transient boiling for conditions slightly above the critical heat flux was carried out to provide a better understanding of the experimental results.

Experimental Equipment

General Description. A schematic representation of the apparatus is shown in Figure 1. This is a modification of the equipment used in Ref. 1, and hence only a brief description of the basic apparatus will be given here, along with a more detailed description of the additions which were made. A cross-shaped platform was constructed which was free to move in the vertical direction. A Fastex motion-picture camera, capable of taking up to 5000 frames per sec, was mounted on the platform along with a rectangular tank (3½ in. × 5⅞ in. × 5½ in. deep) in which the boiling took place.

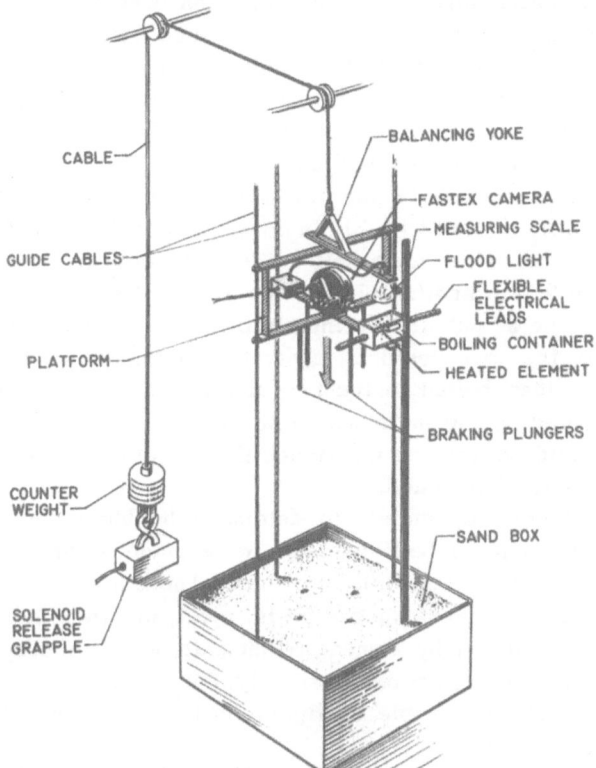

Figure 1. Schematic diagram of counterweighted platform for drop test.

A cable ran from the platform, over pulleys attached to the ceiling, and then down to a counterweight. By changing the size of the counterweight, the downward acceleration of the platform could be varied, and hence experimental runs could be taken at any gravity field in the range between zero and unity. When making a run, the counterweight was pulled down to the floor which raised the platform to the ceiling. The bottom end of the counterweight was grasped by a solenoid operated grapple bolted to the floor. Opening the grapple allowed the platform to fall 9 ft before contacting a sand bed which was used as a braking device. For zero gravity conditions, the platform fell freely after being released from a grapple at the ceiling.

Test sections. An electrically heated element was supported in a brass tank having Lucite windows in the front and back faces. The other surfaces of the tank were covered with insulation to help maintain the water uniformly at the saturation temperature. A copper electrode ½ in in diameter was brought through each end of the tank and supported on the outside by a Teflon flange which insulated it from the tank. Different test sections could be easily clamped to the electrodes. Three types of test sections, which were used for different purposes, will now be described.

The tests to determine the variation of burnout heat flux with gravity were conducted with platinum wires 0.0453 in. in diameter. The ends of the wires were soft soldered to copper blocks which were spaced 2½ in. apart. Voltage taps were drilled into the copper blocks as close as possible to the platinum wire so that the voltage drop across the wire could be accurately determined. The copper blocks were then bolted to the two electrodes. Some of the wires were calibrated by the instrumentation section of the laboratory so that they could be used as resistance thermometers and an average wire temperature determined.

The photographic studies, to determine bubble sizes and rise rates in the nucleate boiling range, were made using flat nickel ribbons, 0.2 in. wide and 0.010 in. thick. These ribbons were also soldered to copper blocks spaced 2½ in. apart, and then their bottom sides were insulated by cementing them to bakelite strips ¼ in. thick. Boiling then occurred from only the top surface, and this provided a relatively simple configuration for making bubble measurements.

For photographic studies of film boiling, the test sections were fabricated from stainless-steel tubing with an outside diameter of

0.157 in. The ends of the tubing were flattened and clamped between copper blocks 2½ in. apart which were bolted to the electrodes. The stainless steel could remain red hot under water for several minutes without burning out.

Power supply. Power was supplied by a 3-phase Mallory rectifier unit which could provide up to 500 amp. d.c. with about 5 per cent ripple. The power was controlled by a system of powerstats on the 220 volt a-c line which supplied the rectifier. The d-c current flow was measured by reading the voltage drop across a 0.001 ohm Leeds and Northrup precision resistor placed in series in the circuit. Large loops of flexible woven welding cable, which could fall freely with the platform, were used to carry the current to the electrodes in the boiling container.

Instrumentation. The voltage across the calibrated resistor, and the voltage across the heated test section were both measured with a Rubicon potentiometer and light beam galvonometer. The current in the circuit could be determined within ± 0.2 amp at 100 amp and the voltage across the test section read within 0.002 volt at 1 volt. Although this instrument was quite sensitive, there were slight current and voltage fluctuations during the boiling process, and hence the heat dissipation could be in error a few per cent. The potentiometer readings were taken just before the equipment was dropped. To read the transient current and voltage while the equipment was descending, a dual-beam oscilloscope (Tektronix Type 535) was used. The bias settings on the two preamplifiers were made to center the initial voltage and current signals, and the gains were set as high as possible so that small changes from the initial values could be observed. Transient temperature changes of the test section were evident if they exceeded about 6°F. While the apparatus fell, the signals made a single sweep at 5 cm/sec and were recorded with a polaroid camera. The 5 per cent ripple from the power supply caused a corresponding ripple in the oscilloscope signals, and this had to be filtered out, since it was of the same order of magnitude as the changes to be observed. This was done with a low band-pass filter which reduced a 60 cy/sec signal by a factor of 20, but allowed a signal with a frequency of less than 12 cy/sec to pass unaffected. For a step-function pulse, the filter gave a full response within 1/20 sec, and hence would adequately follow the signals to be observed. To eliminate spurious signals, the wires in the circuitry were carefully shielded.

The temperature of the boiling water was measured with two copper-constantan thermocouples immersed in the tank and read with a Rubicon potentiometer.

Experimental Procedure

Due to the dependence of the boiling process on many parameters, a great deal of care had to be taken to try to keep conditions the same during successive experimental runs. A learning period of several weeks was necessary before a good experimental technique was achieved. In order to provide a basis for critical evaluation of the work, and to aid any future workers who may use this type of apparatus, a detailed description of a typical burnout test will follow.

Several platinum test sections were fabricated in an identical way, and one was attached between the electrodes in the boiling container. The tank was then filled with boiled distilled water and a 350-watt stainless steel jacketed immersion heater was inserted to bring the water to the saturation temperature. During this time the d-c power was turned on and a moderate current passed through the platinum wire. When the saturation temperature was reached, as indicated by the thermocouples as well as by normal boiling conditions, the immersion heater was removed and the burnout heat flux of the wire was checked. This was accomplished by slowly raising the power in the circuit while watching an ammeter connected into the output of the d-c power supply. At the onset of burnout, the increase in temperature of the heated wire caused an increase in resistance in the circuit and hence a decrease in current. As soon as a reduction in current began, the power was immediately shut off. It was found that sometimes the wire could boil normally for a few minutes at a given power setting and then suddenly burn out; hence the power was raised slowly to provide ample time for the boiling to adjust properly. The transients involved in initiating burnout will be discussed in a later section.

At the instant the wire started to burn out before the power was shut off, the wire would usually begin to glow slightly in one local region, but no physical damage could be observed. However, when the power was restored and boiling resumed, it was noted that the region which had started to burn out was deficient in nucleation centers, possibly due to the high temperature which had been reached at that location. This phenomenon has been observed before by

McAdams *et al.*[6] If the heat flux was immediately brought up again, the burnout flux was much lower, apparently due to the lack of efficient boiling over part of the wire surface. However, if a period of 2 to 5 min was allowed, the nucleation centers would be restored and the burnout flux would return close to its former value.

After the burnout check, several minutes were allowed for the nucleation centers to become re-established. Then the power was set at the desired value for the low-gravity test, and a counterweight corresponding to the required gravity field was placed at the end of the hoisting cable. The platform was then raised to the ceiling and the cable secured by the solenoid grapple. Potentiometer readings of the voltages and current in the test section were recorded, and the thermocouple readings for the water temperature were taken. Upon depressing a switch, the oscilloscope was triggered, initiating a 2-sec sweep and the high-speed motion-picture camera was started to record the initial boiling state before the platform fell. From 0.3 to 0.5 sec later, according to the setting on a timer, the solenoid grapple opened and the platform began to fall. As soon as the drop was complete the power was cut off to prevent the test section from melting in the event that burnout had been initiated while the low-gravity environment existed.

The existence of burnout during the drop was shown by oscillograph traces of the current and voltage. These showed an increase in voltage drop across the heated wire and a decrease in the current flow due to the increase in wire resistance as the temperature increased. Figure 2 shows a typical record of the transient current and voltage when burnout occurred. Depending on whether burnout did or did not occur during the drop, the power was decreased or increased by about 5 per cent and another run taken. This procedure was continued until two runs were obtained, one where burnout occurred, and one at a power about 5 per cent lower where burnout did not occur. After these two runs were achieved, the burnout heat flux was checked again at normal gravity for comparison purposes. This procedure was repeated for several different gravity fields and with several platinum wire test sections.

The effective gravity field for a given counterweight size was determined by measuring the rate of descent of the platform. This was accomplished by placing a stationary vertical measuring scale in the viewing field of the falling motion picture camera to provide a continuous record of platform height on the film. In addition, an argon timing lamp in the camera placed a mark on the film every 1/120 sec. From the data of distance and time, the acceleration of the platform could be determined.

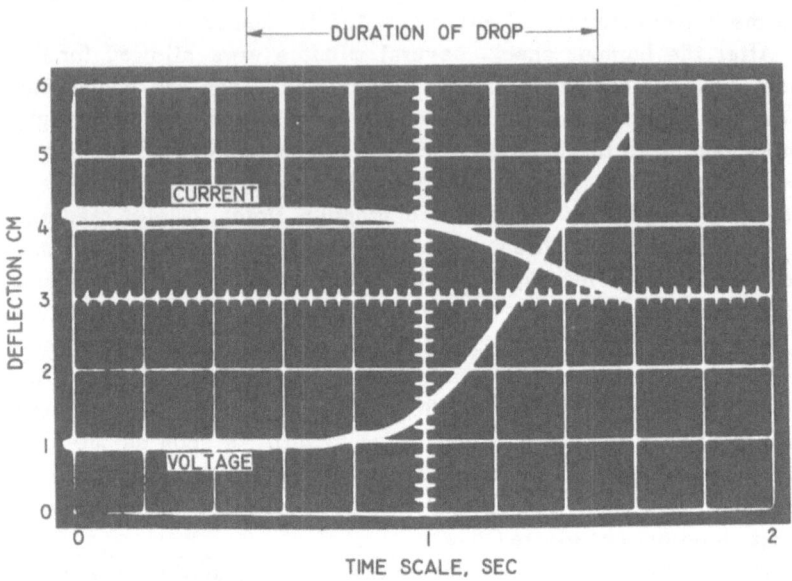

Figure 2. Behavior of voltage and current in a typical drop test with burnout occurring. (Deflection scales: 0.072 cm/amp, 8.5 cm/volt.)

Effect of Gravity on Burnout

For each series of drop tests at a particular gravity, a heat flow per unit area just above and below the apparent burnout point was computed from the current and voltage measurements. These values were then divided by the burnout flux at normal gravity for that particular test section. The effect of gravity on the burnout heat flux is shown by the data on Figure 3. It is seen that there is a considerable scatter in the data at each gravity. This is believed to be due to the transients involved in the burnout process, and some experimental results to explain this will be discussed in the next section. In spite of the scatter, the data does indicate a· definite trend. For example, at 5.5 per cent gravity the burnout flux has been reduced to about 65 per cent of the value at normal gravity. At 47.5

per cent gravity, the wire would not burn out during the drop even for heat fluxes within 7 per cent of the normal gravity burnout limit. Also shown on the curve is a line of $(g/g_n)^{1/4}$ variation. This variation for the critical heat flux has been predicted by a number of investigators as discussed, for example, in (5). The curve appears to provide a reasonable lower limit for the data. The results for zero gravity will be considered in more detail in the next section.

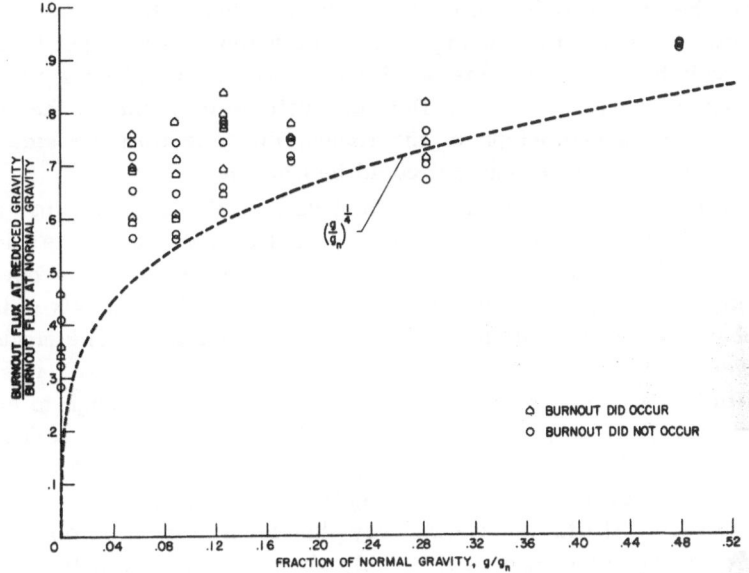

Figure 3. Effect of reduced gravity on burnout heat flux.

Transient Burnout Studies

The data in these experiments were obtained during a time interval on the order of a second while the heated surface was being subjected to a decreased gravity field. The question then arises as to how rapidly the system will respond and indicate a burnout condition after a sudden change in gravity field. For example, if the power were set at the burnout flux corresponding to a certain low gravity, and then the test section were subjected to that gravity, how long would it take for a burnout condition to be indicated? It was necessary to investigate this transient response in order to properly interpret the burnout data.

For this investigation, a resistor and a switch were placed in parallel across the heated wire. When the switch was closed, part of the current would go through the resistor. When the switch was opened, the wire received a step increase in power. Using the same instrumentation as in the drop tests, the experiments were carried out as follows. First the power was slowly raised to determine the burnout flux for the test section. Then the switch on the parallel resistor was closed, and the total current flow was increased so that when the switch was opened, the heat flux in the test section would be a certain percentage above the burnout value. Just before the switch was opened, the oscilloscope was triggered on a 10-sec sweep to record the current flow and voltage drop across the test wire. From a photograph of the traces, the time from the step in power to the onset of burnout could be measured.

In order to obtain good statistical data on the transient time required for burnout, a large number of tests is necessary. The same test section should be repeatedly subjected to the same power pulse until a good distribution curve is obtained which gives the probability for burnout to occur within a given time. The manual power shut-off, used to prevent destruction of the test wire, was a limitation in this experiment, since it was not rapid enough to prevent the wire from glowing slightly red for each test. After a dozen or so tests this would cause changes in the wire, and the burnout limit would decrease so that the data would become inconsistent. Despite this limitation, we can say that for a step in power to 10 per cent above the burnout flux, only about 30 per cent of the burnouts occurred in less than one sec. The remaining 70 per cent took from 1 to 5 sec. However, when the power was pulsed to about 25 per cent above burnout, all of the burnouts took place within one sec and many of them were essentially instantaneous.

These transient tests could only be performed under normal gravity conditions. However, it is reasonable to assume they apply qualitatively for reduced gravity. The results of these tests offer a possible explanation for the scatter which was observed on Figure 3. Suppose that a series of drop tests was being made in which the power was increased in increments until burnout occurred. When the power was just above the burnout flux for the gravity field under consideration, there was a fairly small probability that burnout would occur during the short duration of the test, and hence a burnout would only be observed at this power setting in some instances.

Most likely the power would have been increased and another test performed. Hence, burnout would finally be observed at some higher value where there was a greater probability of the instability taking place within one second. As a consequence the burnout data would generally be too high, but not by over 25 per cent, since at this excess power, burnout within one sec would be expected. The very lowest data points at a given gravity would be near the true burnout limit and would be the result of highly improbable event occurring.

Although there is no experimental evidence on local surface temperature fluctuations in boiling,[3] there is a possibility that these would account for the randomness in the transient burnout behavior. When part of the wire reaches a certain critical temperature, the process becomes unstable and burnout occurs. As the mean temperature level of the wire is raised, the probability is increased that locally the burnout temperature could be exceeded in a given time interval. For a power setting which would cause the average temperature of the entire wire to be above the critical value, burnout would occur almost instantaneously.

The transient considerations are especially important for the zero gravity case. In this instance the theories for the critical heat flux indicate that the burnout flux would go to zero. This of course neglects the heat removal that would actually occur by radiation and and conduction when the heated surface is completely surrounded by vapor. If the liquid is at the saturation temperature, the vapor cavity may keep increasing in size as time proceeds. It is not possible to predict quantitatively what would happen for a long period of zero gravity without performing experiments with longer free fall times than in the present investigation.

Nucleate Boiling at Low Gravity

Photographic study. A photographic study was made at several gravity fields, of boiling from a flat horizontal nickel ribbon with boiling from only the upper surface. The ribbon heat fluxes for the films were all within a few per cent of 91,000 Btu/(hr)(ft^2). The motion pictures were placed in a viewer and measurements were taken on about 15 different bubbles for each gravity field. The number of bubbles measured was limited by the low gravity runs where relatively few bubbles are formed because vapor removal is so poor. As soon as a bubble could be distinguished on the ribbon, measurements were taken, every 1/120 sec, of the height of its center of gravity above the ribbon. These measurements were taken until the bubble had travelled about 1.5 in. upward from the heated surface. The center of gravity was estimated at the center of the area of the bubble when it was attached, and at the intersection of the major and minor axes when the bubble was detached. The values of the major and minor diameters were recorded at about six successive 1/120 sec time increments immediately following breakoff from the surface.

It was found that in the low gravity range, the bubbles would become quite irregular and poorly defined while attached to the surface. This was due to the poor removal of vapor which would cause small bubbles to coalesce and form a larger film of vapor. For this reason, the bubbles attached to the surface were very irregular, and hence an average bubble size for attached bubbles could not be determined. Shortly after the bubbles detached, they became much more distinct, and hence diameter measurements could be made for the detached condition.

Bubble rise rates. Figures 4a and b show how typical curves of the heights of the bubble centers as a function of time for 5.5 and 28.2 per cent gravity fields. Although the bubbles had a fairly wide variation in diameters, it is seen that the rise rates for the detached bubbles are all very much the same for a given gravity field. This is in agreement with the results for high bubble Reynolds numbers[7], where the rise rate of bubbles was found to be independent of size. The bubble paths have some irregularities probably due to the stirring action of the boiling process. The resulting convection currents have increased importance at low gravity fields where the buoyancy forces become small. When looking at the films it was noted that bubbles would move downward at times, due to the convection currents.

Plots similar to those in Figures 4a and b were prepared for each gravity field including the normal one gravity condition. For each bubble path, two straight lines were then drawn. One line went in an average way through all the points after the bubble had detached, and hence its slope gave an average bubble velocity for about one inch of rise away from the heated surface. The other line was drawn tangent to curve of the data at the point of detachment, and the slope of this line gave the bubble velocity at the time of breakoff from the surface. An arithmetic average of each of these two rise rates was then computed using all the bubble paths for each

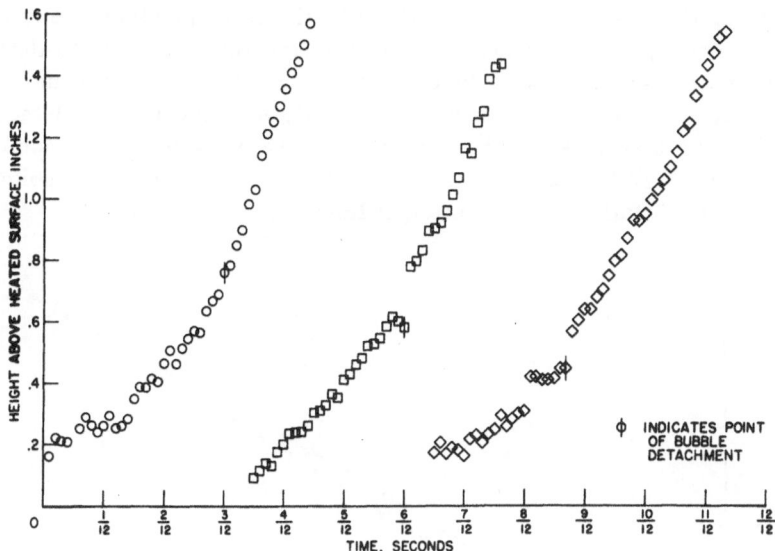

Figure 4a. Rise of individual bubbles in 5.5 per cent of normal gravity.

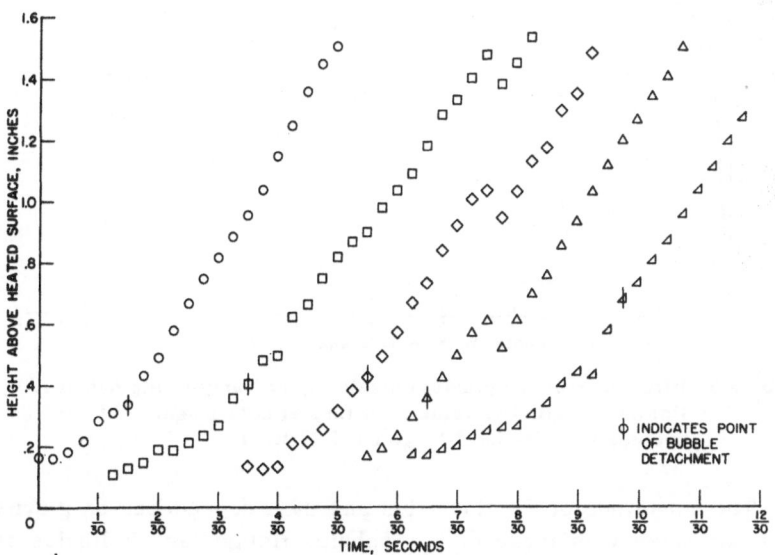

Figure 4b. Rise of individual bubbles in 28.2 per cent of normal gravity.

gravity field, and these were divided by the corresponding rise rates for normal gravity. Figure 5 shows the rise rates relative to those for normal gravity, as a function of the gravity field. For normal gravity the average of the bubble velocities at the time of detachment was 0.98 ft/sec, while the average velocity over one inch of rise was 1.17 ft/sec. Except for very low gravities, the data points on Figure 5 fall close to a straight line.

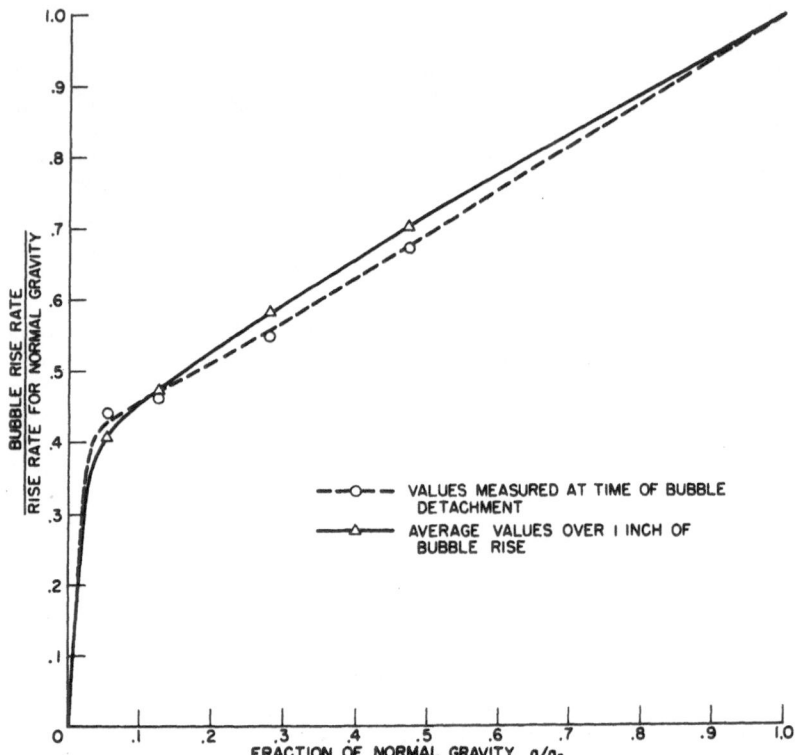

Figure 5. Effect of reduced gravity on velocity of freely rising bubbles. For normal gravity, velocity at time of detachment = 0.98 ft/sec; average velocity for 1-in. rise = 1.17 ft/sec.

The magnitude of the rise rates and their dependence on gravity did not agree with those of gas bubbles rising through liquids as given in Refs. 7 and 8. The rise rate measured for normal gravity was higher than that predicted from the gas bubble correlations.

Even larger rise rates have been observed in Ref. 5. The bubble velocities may have been influenced by cellular convective motions in the tank, which would produce an upward circulation in the vicinity of the heated element, and a downward motion along the sides. In view of these motions, it is not known if the calculation of bubble drag coefficients would be very meaningful.

Bubble diameters. Since most of the bubbles were not spherical, an assumption had to be made to obtain an average bubble diameter from the measurements of the major and minor diameters. Since the third dimension of each bubble into the plane of the photograph was not known, it was decided that any refinement beyond a simple arithmetic average bubble diameter was not warranted. For each bubble, the diameter at detachment was taken as the arithmetic average of several readings taken at successive 1/120 sec intervals immediately following detachment. Then an average was taken of all the bubbles for a given gravity field.

For comparison, information on bubble diameters for the normal gravity condition was needed. This was obtained by starting the motion picture camera about one quarter of a second before the platform was released. The bubble rise rates for normal gravity did not vary appreciably from run to run, but the bubble diameters could vary by several percent. This variation in bubble size was possibly due to small differences in water temperature for different runs. Hence the data on bubble diameter for each gravity field was compared with the normal gravity condition observed immediately before the drop. Figure 6 shows a plot of the ratio of bubble diameter to

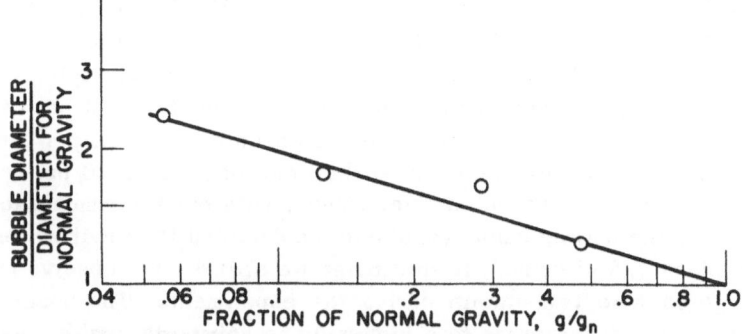

Figure 6. Effect of reduced gravity on diameter of bubbles at time of detachment from surface. Average bubble diameter for normal gravity = 0.245 in.

normal gravity bubble diameter, as a function of gravity. For normal gravity the average bubble diameter for all runs was 0.245 in. As expected, the bubble diameter increases as the gravity field is reduced. The slope of the line on the logarithmic plot is about 1/3.5. A rough comparison of bubble sizes can be seen on Figure 7 which shows typical photographs of bubbles for five different gravity fields with the same heat flux at the surface.

Comments on nucleate boiling data. It was thought that in the nucleate boiling range, a decrease in the gravity field might be reflected in a poorer heat transfer due to the accumulation of vapor near the heated surface. This would require a higher surface temperature to dissipate the same heat flux. Hence, during a drop test, the oscilloscope traces might show a higher voltage drop and a lower current flow due to the resistance increase of the platinum wire. However, in drop tests where burnout was not present the traces went straight across the screen and no surface temperature change could be detected. It should be noted, however, that a change could only be clearly detected if it was at least 6°F, so that a shift in the boiling curve of q/A as a function of $t_s - t_{sat}$ could have taken place of less than this magnitude. These results can be justified in terms of the analysis of Forster and Zuber[3], which yields a heat-transfer coefficient independent of gravity. Their analysis shows that the "stirring action" of the bubbles, as indicated by the product of the bubble radius and the change in radius with time, is independent of the size of the bubbles. Hence the large bubbles obtained in low gravity fields may be just as effective in promoting good heat transfer. The results of Rohsenow[3] do contain a gravity parameter, but it is to a small power. If q/A is constant, then Δt is proportional to $(g_n/g)^{1/6}$. For example, if a drop test is performed where g is reduced suddenly to 20 per cent of normal gravity, then Δt would be increased by about 30 per cent. Since the Δt for these tests is on the order of 25°F, a 30 per cent change would just be detectable. Hence, with our instrumentation, a shift in the boiling curve would only be detected for gravities less than about 1/5 of normal. In this range we still did not observe any change in wire temperature during the experiments. The possible shift of the boiling curve to a higher Δt is unimportant with regard to failure of a heating element by melting, except at very low gravities where a continuous vapor film may be formed. At zero gravity

$g = g_n$

$g = .48 \, g_n$

$g = 28. \, g_n$

$g = .13 \, g_n$

$g = .055 \, g_n$

Figure 7. Nucleate boiling from top surface of a nickel ribbon for five different gravity fields. Heat flux = 91,000 Btu/hr/ft^2; g_n = Earth gravity.

the limitation of the short duration of the test is significant, since under these conditions the surface would eventually tend to become surrounded by vapor.

The experimental data recorded prior to each drop test provide information on normal one gravity nucleate boiling. A boiling curve using these data is presented in Figure 8. The data were taken using

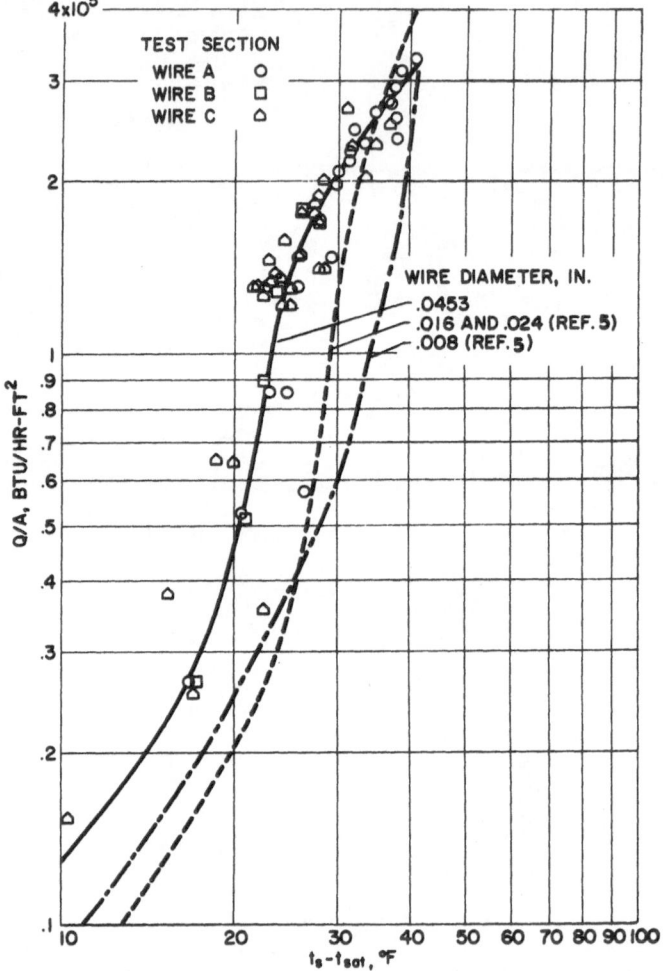

Figure 8. Nucleate boiling from a horizontal platinum wire in water at atmospheric pressure.

three calibrated platinum wires with diameters of 0.0453 in. The resistance measurement of the wire provides an average wire temperature, and a correction for the radial temperature variation was applied as given in Ref. 6 to yield the surface temperature. There was also an axial temperature variation due to heat conduction from the heated wire to the cooler supporting electrodes. This effect was checked using the results of Callendar[9] (p. 152) and found to be quite small, so no correction was needed. The figure also shows how the data are located relative to those for smaller diameter wires given in Ref. (6).

Motion Picture Film

A 16-mm narrated color film (No. C208) has been prepared and is available on loan from the NASA, Lewis Research Center. The film is entitled "Boiling in Reduced Gravity" and was produced by C. M. Usiskin and R. Siegel. The film shows the experimental equipment and illustrates its operation. Nucleate boiling of water from a horizontal platinum wire is shown for three gravity fields; 28, 5.5, and 0 per cent of normal gravity and is compared with normal boiling. The heat flux for these three sequences is 230,000 Btu/(hr)(ft²). Then film boiling is shown from the outside of a horizontal, red-hot stainless steel tube at 5.5 and 0 per cent of normal gravity. The heat flux for these two sequences is 175,000 Btu/(hr) (ft²). The nature of low gravity film boiling is illustrated on Figure 9.

Conclusions

The burnout heat flux decreases with decreased gravity as expected from the trend predicted by theory. The theoretical curve which gives the burnout flux varying as gravity to the one quarter power appears to be a reasonable lower limit for the data. If the heat flux is suddenly raised above the burnout limit an interval of time is required before burnout actually occurs. This transient time decreases as the magnitude of the step in power is increased beyond the burnout limit. Because the data in this work were all obtained during drop tests which only lasted about one sec, the transient phenomenon are significant in accounting for the scatter of the burnout data.

The photographic study of nucleate boiling indicates how the velocity of freely rising bubbles decreases as gravity is reduced.

$g = g_n$

$g = .055 \, g_n$

Figure 9. Film boiling from outside of a stainless steel tube for two
gravity fields. Heat flux = 127,000 Btu/hr/ft^2; g_n = Earth gravity.

The bubble diameters increase as gravity is decreased approximately as gravity to the 1/3.5 power. At zero gravity, film and nucleate boiling take on a similar appearance.

For nucleate boiling, within the limitations of the present experimental equipment, the surface temperature of the test section for a fixed heat dissipation was not influenced by the reduction in gravity field.

Acknowledgments

The authors wish to express sincere thanks for the help and ideas received from other members of the Lewis Research Center. The staff of the photo lab helped us considerably to obtain clear motion pictures. Particular thanks are due to Royal Boyd who fabricated the equipment and assisted during its operation, and Eileen Norris, Marcelle Jordan and Eileen Cox who helped analyze the photographic data.

Nomenclature

A surface area of heated test section
g gravity field
g_n normal earth gravity field
Q total heat dissipation from test section
q heat flux per unit area
t_s surface temperature of test section
t_{sat} saturation temperature of water
Δt temperature difference $t_s - t_{sat}$

References

1. R. Siegel and C. M. Usiskin, "A Photographic Study of Boiling in the Absence of Gravity," ASME Trans., Series C, *Jour. Heat Transfer*, Vol. 81, No. 3, Aug., 1959, pp. 230-236.

2. W. C. Reynolds, "Behavior of Liquids in Free Fall," submitted to Readers Forum, *Jour. Aero/Space Sciences.*

3. J. W. Westwater, "Boiling of Liquids," *Advances in Chemical Engineering*, Volume I, Academic Press Inc., N.Y., 1956, pp. 1-76.

4. N. Zuber, "Hydrodynamic Aspects of Boiling Heat Transfer," thesis, University of California, Los Angeles, June, 1959.

5. R. Cole, "A Photographic Study of Pool Boiling in the Region of the Critical Heat Flux," submitted for publication in *Jour. A.I.Ch.E.*

6. W. H. McAdams, J. N. Addoms, P. M. Rinaldo, and R. S. Day, "Heat Transfer From Single Horizontal Wires to Boiling Water," *Chemical Engineering Progress*, Vol. 44, No. 8, August 1948, pp. 639-646.

7. F. N. Peebles, and H. J. Garber, "Studies on the Motion of Gas Bubbles in Liquids," *Chemical Engineering Progress*, Vol. 49, 1953, p. 88.

8. D. W. Moore, "The Rise of a Gas Bubble in a Viscous Liquid," *Jour. Fluid Mechanics*, Vol. 16, Part I, July 1959, pp. 113-130.

9. H. S. Carslaw, and J. C. Jaeger, *Conduction of Heat in Solids*, 2nd ed., Oxford, Clarendon Press, 1959.

THE EFFECT OF GRAVITY UPON NUCLEATE BOILING HEAT TRANSFER*

MARVIN ADELBERG AND KURT FORSTER

Member, Technical Staff, Space Technology Laboratories, El Segundo California, and Professor of Engineering, University of California, Los Angeles, California

There has been very little experimental work on the effects of acceleration (g) upon nucleate boiling. Two papers [1,2] have reported results for accelerations less than one g. (Usiskin and Siegel [1] explored the range from 0 to 1 g while Steinle [2] reported results for zero g.) Both experiments used drop towers having a free fall of approximately 10 ft corresponding to a time lapse of less than one sec. Thus the results probably suffered from transient effects. Merte and Clark [3] investigated pool boiling for the range 1 to 21 g and did find some influence of g upon boiling. For q/A in excess of 50,000 Btu/hr ft^2 there seemed to be little influence but for lower heat fluxes a definite trend was found.

There are two generally accepted analysis which rely, at least in part, upon theory; the Forster and coworkers theory and that of Rohsenow and coworkers. These two semitheoretical results are presented in the form of

$$N_u = CR_e{}^{n_1} P_r{}^{n_2}$$

In the Rohsenow formulation the term g appears while it is not present in that of Forster and coworkers. Usiskin and Siegel did not find any influence of g upon nucleate boiling and suggested that the correlations of Forster and co-workers have been substantiated. For reference, the two correlations mentioned above are presented below:

$$\frac{h\beta}{k_L}\left[\frac{g_0\,\sigma}{g\,(\rho_L - \rho_V)}\right]^{1/2} = \tag{2}$$

$$(\text{Const.})\left[\left(\frac{\beta q/A}{\mu_L\lambda}\right)\left(\frac{g_0\,\sigma}{g\,(\rho_L - \rho_V)}\right)^{1/2}\right]^{2/3}\left(\frac{C_L\mu_L}{k_L}\right)^{-0.7}$$

---

*This paper was not formally included in the original program; however its contents were delivered by Dr. Adelberg and Professor Forster in the form of comments and discussions during the course of the Symposium and the following Panel on "Technical Applications of Zero-G Research."

$$\frac{C_L \rho_L \sqrt{\pi a_L}\, q/A}{k_L \rho_v \lambda} \left(\frac{2\sigma}{\Delta p}\right)^{1/2} \left(\frac{\rho_L}{\Delta p}\right)^{1/4} =$$

$$= 0.0015 \left[\left(\frac{\rho_L}{\mu_L}\right)\left(\frac{C_L \rho_L \Delta T \sqrt{\pi a_L}}{\rho_v \lambda}\right)^2\right]^{0.62} \left(\frac{C_L \mu_L}{k_L}\right)^{1/3} \tag{3}$$

We wish to note, however, that the fact that g does not appear *explicitly* in the correlations of Forster and coworkers does not imply that this correlation is independent of g. The correlation constant C in that correlation was obtained by reference to experiments performed under normal gravity (g = 1); for example, had C for the correlation been determined by reference to experiments performed on the moon ($g = 1/6$), then the influence of g would have been manifested in a new value of C.

To elaborate on this point, by reasoning from dimensional analysis: assuming four primary dimensions (time, temperature, mass and length) and noting the large number of significant and independent variables which appear in the various correlations, it is evident that more than three π terms are required for a complete set. Thus the additional π terms which do not appear explicitly are incorporated in the constant C. Should g be a significant independent variable, then at least one of these additional π terms in C contain g.

We want to take this opportunity to present some physical considerations concerning the relative importance of gravity in the transfer of heat from a heated surface to a boiling liquid. The main function of gravity in this process is the removal of the vapor bubbles from the heated surface. Apart from gravity, the vapor bubbles can be removed from the surface also by forced convection (shear forces) and/or by the flow field induced by the dynamics of the growing bubble. Let us estimate the dynamic force F_d associated with the growing bubble and compare it with the buoyancy force F_b in order to obtain a criterion for the relative importance of the latter. Consider a hemispherical cavity of radius $R(t)$ growing on a flat surface in a liquid of density ρ; its apparent mass is of the order $2/3\rho R^3 \pi$, and its velocity is of the order $\dot{R}$. We set the dynamic force F_d associated with this mass flow proportional to the momentum change

$$F_d \sim \frac{d}{dt}(2/3\rho\pi R^3\dot{R}) = 2\rho\pi(R^2\dot{R}^2 + 1/3R^3\ddot{R}) \tag{4}$$

and we submit that in a system in which the buoyancy force dominates, the ratio F_d/F_b is small and that when this ratio is large, the system will be essentially independent of gravity as far as removal of vapor from the heating surface is concerned. We therefore form this ratio, which is of the nature of a Froude number, π,

$$\pi = \frac{2\rho\pi(R^2\dot{R}^2 + 1/3R^3\ddot{R})}{2/3\rho\pi R^3 g} \tag{5}$$

It is well known that for a bubble growing in either a saturated or a subcooled liquid, $\ddot{R}$ is negative and inasmuch as the present is a consideration in orders of magnitude, we neglect the second term in Eq. (2) and we approximate $\dot{R}$ by the mean value theorem

$$\dot{R} \approx \frac{R_{max}}{t_{max}} \tag{6}$$

in terms of the maximum radius R_{max} and the time t_{max} at which this maximum radius occurs. Hence

$$\pi = \frac{3R_{max}}{g t^2_{max}} \tag{7}$$

The problem for the designing engineer is to develop a system under normal ($g = 1$) conditions such that he can predict its performance for other gravity levels. Equation (4) indicates that the important ratio is that of R/t^2 versus g. For saturated boiling at $g = 1$, significant times for detachment of the vapor bubbles are of the order of 3×10^{-2} seconds and the corresponding maximum radius about 0.4 cm. For this condition π, from Eq. (4), is seen to be about unity. Without resorting to new experiments, we can take advantage of the fact that for a subcooled liquid, in which the bubbles collapse due to condensation (rather than being detached by gravity), the boiling mechanism should not be significantly influenced by gravity. We might conclude from this that values of π corresponding to the subcooled case indicate a system which is insensitive to changing conditions of gravity. To obtain a feeling for the magnitude of π

under such conditions, we tabulate below, some experimental data
reported by Ellion[4]:

TABLE 1: EFFECT OF SUBCOOLING UPON THE RATIO π

Liquid Temp., $T_L - {}^\circ F$	R_{max} cm $\times 10^2$	t_{max} sec $\times 10^{-3}$	π (Eq. 4)
212	40	30	1.3
170	5.3	1.1	130
120	4.4	0.83	190
60	3.4	0.50	410

We conclude that a system which is characterized at $g = 1$ by a
value for π greater than, say, 100 is essentially insensitive to
change of gravity. The relatively very low value of π in the first
row of the above table suggests that a system designed for pool
boiling of a saturated liquid at $g = 1$ will fail at low g levels. This
is so for the following reason: in order to produce sizable dynamic
forces of the growing vapor bubbles in a saturated liquid, the superheat would have to be extremely high; but superheat is limited to
low values by the occurrence of burnout. Deeper insight into these
questions might be obtained by the experiment suggested below.

Some experiments were suggested by one of us a few years ago
which were intended to yield design criteria for heat transfer components to operate independently of the level of gravity. Essentially,
the designer needs to know two criteria regarding the boiling process, for his system designed under $g = 1$ conditions, to remain
largely unaffected by reduction in gravity: (1) The level of subcooling and/or superheat at which the bubble *dynamic forces* dominate and are sufficient to insure removal of the bubbles from the
heating surface; (2) The velocity in forced convection at which the
shear forces near the heating surface dominate and are sufficient to
remove the vapor from the heating surface, A simple experiment that
will yield the above design criteria and permit the observation of
the phenomena in the steady state is the following: Consider a
system with a flat heating surface on top of and in contact with the
liquid. In this case, gravity tends to keep the bubbles attached to
the heating surface instead of aiding their removal (as takes place
in the "heating from below" condition); in this manner, one can
observe a system in which g is effectively -1, and one can obtain

experimentally, the relation between boiling heat transfer and the usual parameters (superheat, subcooling, convective velocity, etc.). The desired above-mentioned criteria would be established for the range $-1 < g < +1$ when superheat, subcooling and convective velocities were found such that the resulting heat flux in this "heating from above" case approximates the heat flux obtained by "heating from below." In this manner, criteria for sufficient conditions for proper heat transfer design at low gravity can be established by observing the operation at $g = -1$.

In a private communication, Usiskin and Siegel have furnished experimentally determined data necessary to evaluate π (Table 2).

TABLE 2: DATA TO EVALUATE π

Fractional gravity, g/g_n	R_{max}, in.	t_{max}, sec	π $3R_{max}/gt^2$ max
0.055	0.29	0.32	0.40
.126	.22	.17	.47
.282	.20	.14	.28
.475	.15	.067	.55
1.0	.12	.045	.46

They point out that the π term does not exhibit any definite trend with gravity and remains essentially constant. Inasmuch as they found no significant effect of gravity on nucleate boiling heat transfer either, the constancy of the above Froude number is consistent with their experimental findings.

References

1. C. Usiskin and R. Siegel, "An Experimental Study of Boiling in Reduced and Zero Gravity Fields," page 75 of this volume.

2. H. F. Steinle, "An Experimental Study of the Transition from Nucleate to Film Boiling Under Zero Gravity Conditions," *Proc. 1960 Heat Transfer and Fluid Mechanics Inst.*, Stanford University, June, 1960, pp. 208-219.

3. J. A. Clark and H. Merte, Jr., "Pool Boiling in an Accelerating System," A.S.M.E. Paper No. 60-HT-22.

4. Max E. Ellion, "A Study of the Mechanism of Boiling Heat Transfer," Jet Propulsion Laboratory Memorandum No. 20-88, California Institute of Technology, March 1, 1954.

LIQUID BEHAVIOR INVESTIGATIONS UNDER ZERO AND LOW GRAVITY CONDITIONS

E. W. SCHWARTZ

Research Group Engineer, Convair (Astronautics) Division, General Dynamics Corporation, San Diego, California

The influence of a variable gravity field upon liquid behavior phenomena has been the subject of periodic studies at Convair-Astronautics for several years. Initially these efforts centered about means for increasing maximum boiling heat-transfer rates, particularly with respect to water. However, as interest in space missions increased, many of these efforts were redirected toward the heat-transfer and liquid behavior conditions anticipated under zero gravity. Zero gravity, or simply zero g, may be defined in a generalized way as the condition experienced in orbital motion by relatively low-mass bodies such as artificial satellites. In zero g, gravity is not absent; rather, the only resistance to gravity's effects is that due to inertia. Thus it differs from the more basic condition of no gravity, which will probably be approximated only in intergalactic space. This discussion will deal with zero g, and no g can be left for future speculations and explorations. Our immediate interest in zero g was the mechanism of heat transfer to cryogenic liquid propellants that might be employed in those space vehicles requiring a long coasting period. As we got into the program in support of the Centaur Project for the National Aeronautics and Space Administration, it became evident that other zero g problems existed, all of which were important to space vehicles. An exploratory test program to ascertain the problems peculiar to liquids in orbit was initiated early in 1959. The problems considered were:

(a) Liquid behavior
(b) Heat transfer
(c) Venting
(d) Pumping
(e) Settling under a low g condition

An agreement was then made with Wright Air Development Division, Dayton, Ohio to participate with them in zero g flights in their C131B airplane. The C131B airplane is capable of giving zero g conditions for periods up to 15 sec duration. This is accomplished at an altitude of about 12,000 ft by first putting the airplane in a shallow 10 deg dive until the maximum allowable airspeed of 250 knots is reached as shown in Figure 1. At this point a 2½-g pull-up is executed until a flight path heading of about 35 deg above the horizon is reached. Then the airplane is flown in a no-lift, thrust-equal-to-drag, path. This orbital path is terminated when the airplane again reaches its maximum allowable airspeed of 250 knots. At this time it is headed downward in a path of 35 deg below the horizon. The summarized results of the exploratory tests follow.

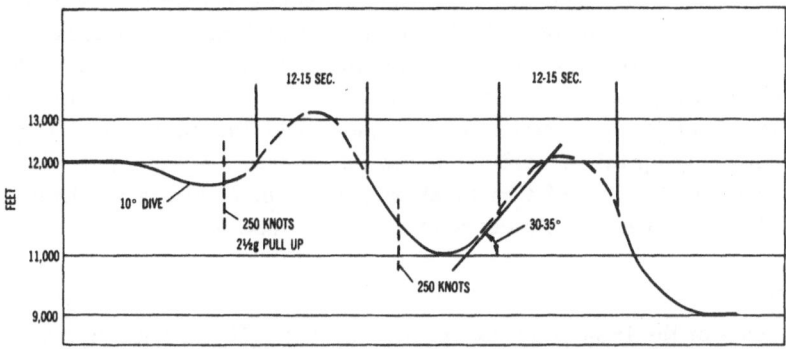

Figure 1. Maneuver of C131B airplane for production of zero gravity conditions.

Liquid Behavior

During the testing, observations were made using plastic tanks or double-walled, unsilvered, glass Dewars. Liquids that normally wet surfaces, such as container walls, proceed to wet those surfaces that had not been previously wetted due to the restraint imposed by resistance to gravity. Under zero g the ullage volume was found to be centrally located within a container whose walls were wetted by the liquid. Residual inertia effects, unequalized pressure forces, and wettable surfaces other than the container walls would upset this balance, however. From the visual observations, the liquid would form a pronounced meniscus at the wetted wall surface as the flight path approached closely to the zero g condition. It is evident that surface tension forces predominate in the near zero g and the zero g condition.

Residual g and inertial forces that momentarily interrupted the zero g condition tended to form bubbles in the liquids. These bubbles did not exhibit a tendency to coalesce during the zero g durations of these tests.

Cryogenic liquids did not behave differently than noncryogenic liquids in the same realm of density, viscosity, and surface tension. However, cryogenic liquids are more prone to wet nearly all surfaces than are noncryogenic liquids.

Heat Transfer

Attempts were made to obtain heat-transfer measurements during the testing. It was anticipated that a transition from nucleate to film boiling would occur on a continuously heated surface under zero g. A heated wire apparatus was set up to confirm this supposition. However, convection of the fluid past the wire was always present during initial testing, which obliterated the expected effect. Visual observation of film blanketing was not observed in airplane flights until very stable conditions were established.

Venting

One of the immediate consequences of the flights with cryogenic liquids was the urgent requirement to vent shortly after entering the zero g condition. This requirement occurs because the upper portion of the Dewar and its cover attain a temperature substantially higher than that of the cryogenic fluid during the period prior to the zero g trajectory. The cryogenic fluid, due to wetting and inertial forces, is brought into contact with these hotter surfaces when the airplane goes into orbit. The boil-off that results often caused a pressure rise that reached 10 to 15 psi within the Dewar.

Material vented to reduce the internal pressure in the Dewar consisted of both gas and liquid. As it is desirable to vent only gas, various schemes were tried, such as the employment of non-wettable surfaces, complex baffle arrangements, and strategic placement of the vent inlet, in an attempt to effect differential venting. These passive arrangements were not successful in practice as the zero g conditions were not ideal during the testing. Similar disturbances to motion might be expected in space vehicles, especially in the immediate period following cessation of thrust.

Various mechanical devices were then tried. These included, (1) heat exchangers, in which the vented liquid is boiled within the unit at a pressure below that existant in the tank being vented, (2) and centrifugal separating devices to produce an artificial gravity field. In the heat exchanger, the lowered boiling temperature theoretically permits an extraction of heat from the fluid in the container being vented. This extraction of heat is beneficial, as it reduces the magnitude of the problem at its source. However, the heat exchangers tested required heat transfer under zero g conditions, an uncertain process at present unless forced convection is present. Therefore, heat exchangers to date were not fully successful, possibly due to the lack of sophistication in the apparatus.

Centrifugal separators have proven to be effective in separating liquid from gas phases once initial rotating seal difficulties were overcome. Our most successful type to date has been designated as a radial inflow device because the vented vapor must take an inward flow path.

Pumping

Pumping under zero g conditions presented no problems not heretofore encountered. Pumps operated normally when the inlets happened to be immersed within the liquid. During those intervals when only vapor was present at the pump inlets, pumping ceased. It was observed that cavitation occurred more readily under zero g conditions as the liquid head, normally present due to pump location, is absent during zero g. Greater problems were encountered in the airplane as engine oil pumps often failed to maintain pressure throughout the zero g trajectory. WADD personnel solved this problem with expulsion bladders on the inlet side of the oil pumps.

Settling Under a Low G Condition

After a long period of coasting in orbit, liquid propellants must be collected and the pump inlet covered to permit reliable operation. It seems obvious that the best method to effect collection is through the utilization of an acceleration imposed by an auxiliary rocket system not influenced by zero g.

Following the coasting period observations and studies show that liquid propellants may be found in one of the following conditions:

(a) A fog
(b) A bubblelike, soap suds type mass
(c) Adhering to all walls with ullage space centrally located
(d) Floating as one or more almost shapeless masses within the tank

Conditions (a) and (b) would be expected following a period of considerable agitation. Conditions (b) and (c) have been observed in flight tests. Condition (d) would be met when the propellant did not wet the wall, or possibly as a result of film boiling in orbit.

Current testing is not adequate to formulate the limiting parameters fully. Nevertheless, qualitative tests and theoretical analyses have effectively shown that settling can be accomplished for all cases.

Test Techniques

Initial exploratory testing in the C131B airplane were conclusive in demonstrating that we have problems in handling liquid propellants in zero g. It was also clear that, despite the very skilled piloting of the aircraft, small residual g forces were always present. It would be highly desirable to eliminate these residuals in the testing. Therefore, in planning a second series of tests we arrived at a new concept in testing technique which we designated as the "free floating" package (see Figure 2).

This package consisted of an instrumented experiment suspended by strong cords within a cagelike enclosure for safety reasons. When the airplane entered the zero g condition, the suspension lines were loosened by the operator who controlled motor-actuated multiple screw-jack mechanisms. Thus loosened, the package could "float" about 18 in. in any direction at right angles to the flight path. By fitting the airplane with rails, using guided rollers on the suspension cage, and employing a reversible winch, a total fore-and-aft "float" of the package of 19 ft could be accommodated.

It was hoped that the package would enter its elliptical, orbital path and remain in it for the entire zero g airplane trajectory free of the inevitable disturbances. Testing showed that this arrangement

Figure 2. Restrained free-floating package for airborne production of zero
gravity conditions in liquids.

was indeed capable of giving better zero g conditions than possible
with the bolted-to-the-airplane configuration. However, it also had
limitations. The elasticity in the suspension cords stored considerable energy during the 2½-g pull-up.

This energy was released to the package during the onset of
zero g. The result was two separate but closely related orbits — one
by the airplane and the other by the zero g package.

The divergence of the separate path would, of course, be interrupted when the limits of the restraining cords or cage rails were
reached. Interruption of the path usually occurred after only 5 to 6
sec had elapsed. Hence, the technique was only partially successful. However, the desirability of the free-floating concept was effectively demonstrated.

Several experiments were then performed by WADD and Convair-
Astronautics personnel to extend the free-floating concept. The use
of the entire airplane cabin was deemed feasible as long as protection to both package and airplane was provided for the inevitable

impact with each other upon resumption of the high g force during
pull-out maneuver. After suitably protecting the airplane with mat-
tresses, normally used by WADD during zero g Aero-Medical experiments, this modified free-floating technique was tried (see Figure
3). Excellent results were obtained. Periods up to 10 sec of very
good zero g were observed in the C131B airplane.

Figure 3. Unrestrained free-floating package for airborne production of
zero gravity conditions in liquids.

One technique employed to effect a constant low g settling
acceleration on the free-floating package is particularly interesting.
It consisted of a long single strand of model airplane rubber-motor
freely stretched between the pilot's cockpit and the experiment at
the aft end of the fuselage. It was found that the force exerted by
the single rubber strand was relatively constant over a reasonable
fore-and-aft displacement of the experimental package. Furthermore,
the thrust of the airplane could be made slightly positive and virtu-
ally eliminate fore-and-aft displacements. The magnitude of force

exerted was that corresponding to approximately .01 g, although other values could also be obtained. The duration of the settling experiments equalled the free-floating period.

During these flights it was also learned that the airplane's intercom system was a definite asset in advising the pilots of the position of the package in the airplane from time to time. The pilots were able to make some airplane flight-path corrections to fit that particular orbit into which the package happened to have been launched. This extended the possible zero g "free floating" period.

Based upon this success, a new series of tests have been planned around the free-floating concept. It is desirable that the pilots know at all times the path of the package and its position in the airplane to permit airplane flight path correction. Analysis of the problem revealed that a commercial closed-circuit television system should serve adequately for this purpose.

As the C131B airplane is limited to a zero g path of 15 sec, alternate airplanes were studied by WADD, NASA, and Convair-Astronautics. A decision was reached to use the KC135, a jet tanker, for this purpose. WADD then acquired a KC135, and has refitted it for zero g testing. A considerable number of zero g trajectories have been completed by this airplane. The total zero g time in each trajectory run has averaged 32 sec (Figure 4). Prior to the installation of the closed-circuit TV, free-floating experiments up to 16-sec duration were accomplished. Longer periods are anticipated with TV.

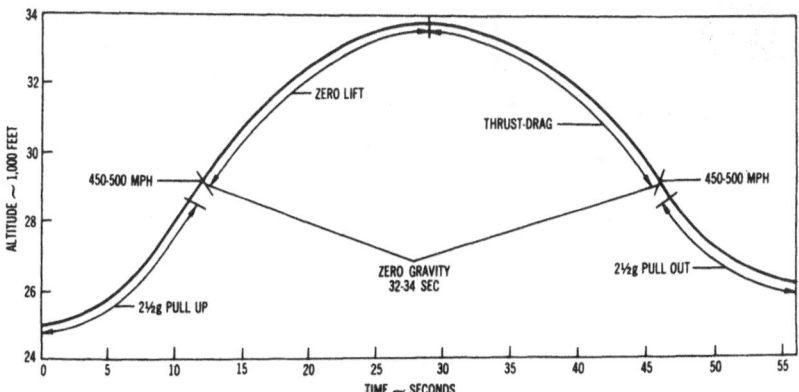

Figure 4. Maneuver of KC-135 jet tanker for production of zero gravity conditions.

Conclusions

As the result of the current experimental program and analyses, it is evident that the employment of airplanes as zero g laboratories is not only feasible, but also affords a method for readily obtaining orbital conditions repeatedly in which experiments of various kinds can be conducted and observed.

The free-floating concept appears to offer the best test method for obtaining orbital zero g conditions in an airplane.

AN EXPERIMENTAL STUDY OF THE TRANSITION FROM NUCLEATE TO FILM BOILING UNDER ZERO GRAVITY CONDITIONS*

HANS F. STEINLE

Thermodynamics Engineer, Aerophysics Group,
Convair (Astronautics) Division, General Dynamics Corporation,
San Diego, California

The handling and control of fluids in satellites and advanced space-vehicle development requires extensive theoretical and experimental investigations in the low-gravity field. Liquid behavior and heat transfer studies are of special interest and some studies have already been performed,[1,2,3,4] but all experimental investigations are of qualitative nature. The purpose of this study is to initiate a quantitative experiment and to demonstrate the sensitivity of the transition point from nucleate to film boiling (maximum heat flux, q/A max) to the gravitational acceleration. This can theoretically be shown by varying the g in the equations derived by Zuber or Kutateladze-Borishanskii[5] which predict q/A max proportional to the 0.25-power of the gravitational acceleration, which means an increase of the heat-transfer rates at q/A max with increasing g, and a decrease with decreasing g.

To perform this study, it was first necessary to establish a method that would definitely indicate whether the energized heating wire submerged in the liquid was only producing nucleate or was actually causing film boiling. This method should also allow precise voltage or current measurement over the short zero g period for heat-transfer studies. The slow response time of indicating instruments led to the introduction of an oscilloscope to record the temperature history of the heater wire. This recording method, which can be applied to the study of boiling processes in high- or low-gravity fields, was found to be highly effective and could still become more refined.

Reprinted from Proceedings of the 1960 Heat Transfer and Fluid Mechanics Institute with the permission of the publishers, Stanford University Press. Copyright 1960 by the Board of Trustees of the Leland Stanford Junior University.

Freon 114 was used as test liquid because of its high boiling point and its low heat of vaporization. It could be maintained quiescent without excessive convection currents in a glass Dewar flask. A comparison of physical properties between water and Freon 114 is shown in Table 4.

Test Apparatus

A relatively simple experimental setup was used (Figure 1). The Dewar flask with Freon was supported by a drop platform and clamped in position by a lid. Four vertical steel rods were fastened

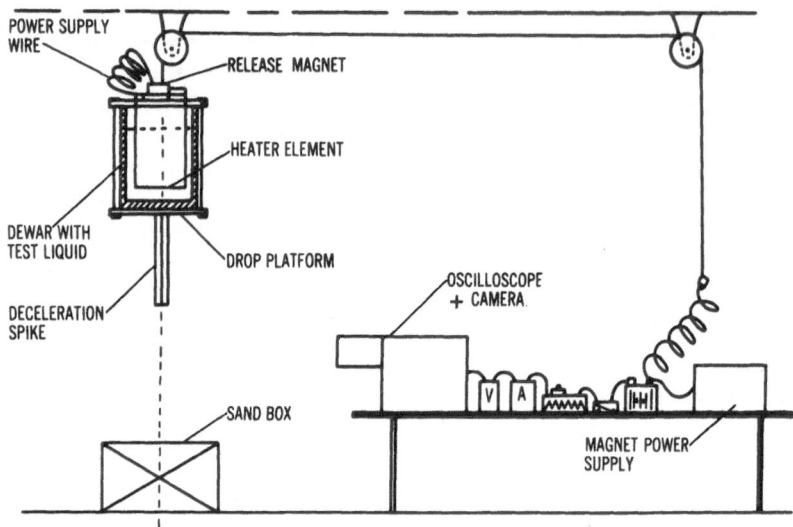

Figure 1. Method for study of the transition from nucleate to film boiling in a zero gravity environment.

to the bottom of the platform for deceleration while plunging into a box of sand. This method was introduced previously by Siegel and Usiskin.[1] A 0.0015-in. diameter, 1.281-in. long platinum wire, suspended between two rods inserted into the Dewar lid, was chosen as the heater element. An electromagnet released the platform after it was raised 9 ft to the laboratory ceiling by a pulley system. The heater voltage was supplied by a group of silver-cell batteries, through trailing wires, and was applied immediately upon release of the platform. Several carbon-pile resistors were connected in series with the batteries and heating wire to provide fine adjustment and constant current. The battery voltage was stabilized by use of a dummy load prior to a drop. The sweep on the oscillograph was triggered by the voltage across the heater wire and recorded by an oscillograph recording camera.

Test Procedure

A major problem was the attainment of reproducibility of the wire-temperature history under apparently identical test conditions. Reproducibility was found to be influenced mainly by two factors: nucleation centers on the heater surface, and convection currents in the liquid. Both problems were solved by the use of controlled wire contamination and by keeping the Freon in a Dewar flask. Controlled contamination refers to covering the heater wire uniformly with nucleation centers which are the precondition for bubble generation.

A series of drop tests was performed. Each test was considered as having three phases: the predrop, the drop, and the postdrop. The current level was set at the beginning of each test and remained constant throughout all phases of the test. The voltage across the heating wire was recorded as a function of time for each phase of the test. The voltage record was made by a single sweep of the oscilloscope trace which was initiated at the instant the heating wire current was switched on. The sweep durations were adjusted to 0.75 sec so that recording was extended over one complete drop period. Thus, there were three measurements or traces made for each test: the first under 1-g conditions, just prior to the drop; the second during the drop; and the third, again under 1-g, immediately after the drop. The current was increased, for succeeding tests, up to the point at which transition from nucleate to film boiling would occur under 1 g.

All tests were performed after the heating wire had been immersed in clear water for three to five min prior to immersion in the Freon. This procedure, used for controlled contamination, gave good repeatability. However, once the wire produced film boiling in the Freon, it had to be recontaminated.

Either nucleate or film boiling could be obtained under the same initially fixed current setting when the setting was selected so as to cause the heat flux to be in the neighborhood of the transition point.

Data Analysis and Discussion of Results

Typical oscillographic pictures are shown in Figure 2, a comparison of oscillographic pictures obtained in 1-g nucleate and film boiling, and zero g boiling for the same current setting. The voltage across the heating wire, which is proportional to the wire temperature in the observed region, was recorded as a function of time.

The curves rise, under 1-g conditions, in the region of natural convection and film boiling under the heating rates and for the time period used in this study. This is due to the fact that more heat is generated than can be carried away through the liquid, and the time period has not permitted equilibrium to become established. A situation where the temperature rise may be sufficient to cause burnout of the heating element can occur in film boiling, where the heat has to be transferred through a vapor film by conduction and radiation. A different situation occurs in nucleate boiling, where vapor bubbles break away from the surface of the heating element and rise through the liquid. For this case, the heat-transfer coefficients are generally very much larger than those in the case of film boiling. A falling curve can be observed in the transition region between natural convection and nucleate boiling in the pictures obtained in 1-g nucleate boiling (Figure 2a), which indicates that the heating wire cools down. A possible explanation of this phenomenon is the fact that the liquid is forced to higher velocity around the wire due to bubble generation causing increased heat transfer. This explanation would be in agreement with the observation that the wire cooldown does not occur in film boiling where the fast generation of a film after a relatively short nucleate period does not allow increased heat transfer (Figure 2b).

The curves at zero g look quite different from those recorded at 1 g (Figure 2c). The region of natural convection is nearly eliminated. Consequently, boiling starts sooner as well as at a lower heating rate than at 1 g. This is due to the fact that the heat is largely stored around the heating element because buoyant forces in the liquid are absent. Nucleate boiling is present only for a few milliseconds before a rising curve indicates decreasing heat transfer caused by a conglomeration of bubbles or a vapor film formed around the heating wire. This type of boiling is referred to as "zero g film boiling," since its nature is not known. Generally, for the same current setting, zero g film boiling starts sooner than film boiling

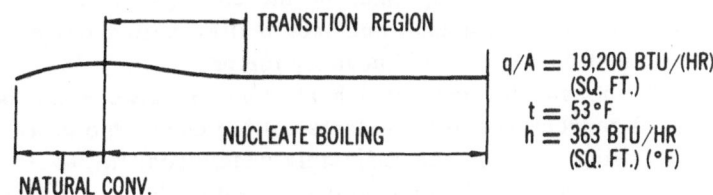

A. NUCLEATE BOILING ONE g

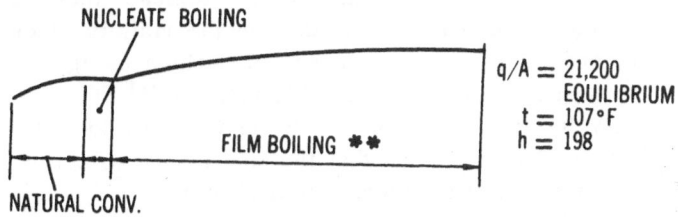

B. FILM BOILING ONE g

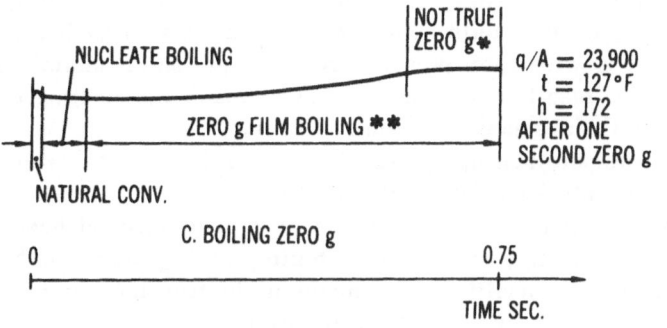

NOTE: *CAUSED BY AIR DRAG
**INCL. TRANSITION BOILING
CURRENT: 8 AMPS

Figure 2. Typical oscillograms of voltage across heating wire.

at 1 g. This can be explained by the assumption that the bubbles generated in the previous nucleate boiling period coalesce faster because of the absence of buoyancy forces.

The reason for increased heat transfer between the period of natural convection and of nucleate boiling under zero g, as recorded by a falling curve, is not completely understood. However, it is felt that the decreasing heating-wire temperature in this region may be due to an elevated heat-transfer coefficient caused by liquid agitation when bubble formation is initiated. This phenomenon does not seem to be typical for zero g, since it was observed also in nucleate boiling at 1 g. Little is known about possible bubble movement in boiling at zero g. Photographic studies indicate that "during the free fall the vapor bubbles remained in the vicinity of the heating surface, and there was no evidence of bubbles being pushed away from the surface with any appreciable velocity."[1] The zero g curves, especially for lower currents, have the tendency to fall after 0.50 to 0.60 sec. This indicates that the true zero g condition does not exist near the end of the drop, due to air drag.

The following general conclusions can be made by comparing the oscillographic pictures and observing the type of boiling before and after the drop:

(a) With the heating rates used in this study, boiling under zero g conditions starts sooner as well as at a lower heating rate than at 1 g, with no or only a very small period of natural convection, as a result of the heat storage around the heating element following the absence of buoyancy forces.

(b) Nucleate boiling appears to be time dependent at zero g and is present only for a few milliseconds.

(c) Zero g film boiling occurs with repeatability at heat fluxes which would permit only nucleate boiling at 1 g, from which it follows that the transition from nucleate to film boiling is clearly sensitive to the gravitational acceleration.

(d) Due to the time dependence of the nucleate period, no clear boiling transition point from nucleate to film boiling was found under zero g conditions.

(e) The conglomeration of bubbles collected or the film formed around the heating element at zero g dispersed very quickly upon return to normal gravity, with the result that the heat transfer phenomena also return to normal.

The quantitative results obtained in the study are shown in Tables 1 to 3 and are summarized in Figure 3 as heat flux versus temperature difference between heating wire and liquid. Since the main objective of the experiment was to study the transition from nucleate to film boiling, efforts to obtain a high degree of quantitative accuracy was not emphasized but was set aside as one of the objectives of forthcoming work.

TABLE 1: DATA ON NUCLEATE BOILING UNDER 1-G CONDITIONS, CURRENT APPLIED SUDDENLY, EQUILIBRIUM ESTABLISHED

I Amp	E Volts	R Ohms	q/A Btu/(hr) (sq ft)	Δt °F	h Btu/(hr) (sq ft)(°F)	Remarks
7.2	0.255	0.03542	14,900	44	340	
7.8	0.280	0.03590	17,800	51	347	
8.1	0.292	0.03605	19,200	53	363	
8.3	0.300	0.03615	20,200	55	370	
9.0	0.328	0.03644	24,000	60	400	
9.4	0.344	0.03660	26,300	62	425	Peak heat flux

The current was increased till film boiling occurred at $q/A > 26,300$.

TABLE 2: DATA ON FILM BOILING UNDER 1-G CONDITIONS, CURRENT APPLIED SUDDENLY, EQUILIBRIUM ESTABLISHED

I Amp	E Volts	R Ohms	q/A Btu/(hr) (sq ft)	Δt °F	h Btu/(hr) (sq ft)(°F)	Remarks
7.8	0.50	0.0641	31,700	525	60.4	Minimum heat flux
8.1	0.55	0.0679	36,200	591	61.3	
8.3	0.58	0.0698	39,100	625	62.6	
9.0	0.70	0.0777	51,200	768	66.7	
9.4	0.77	0.0819	58,800	845	69.6	

TABLE 3: DATA ON BOILING AFTER 1 SEC ZERO GRAVITY CONDITION, CURRENT APPLIED SUDDENLY

I Amp	E Volts	R Ohms	q/A Btu/(hr) (sq ft)	Δt °F	h Btu/(hr) (sq ft)(°F)
7.2	0.280	0.03888	16,400	100	165
7.8	0.318	0.04077	20,200	134	151
8.1	0.363	0.04481	23,900	197	121
8.3	0.370	0.04457	25,000	193	129
9.0	0.448	0.04978	32,800	280	117
9.4	0.420	0.04468	32,100	195	165

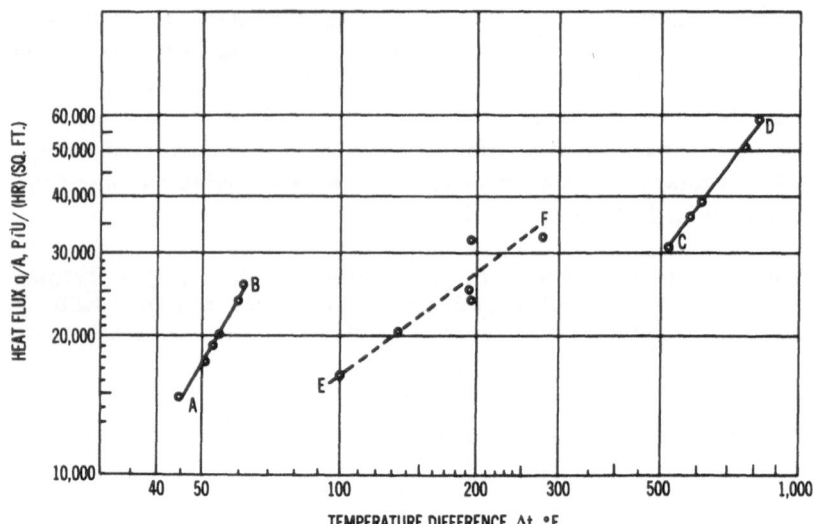

LEGEND:
A-B   NUCLEATE BOILING ONE g, EQUILIBRIUM ESTABLISHED
 B   PEAK HEAT FLUX
C-D   FILM BOILING ONE g, EQUILIBRIUM ESTABLISHED
 C   MINIMUM HEAT FLUX
E-F   BOILING ZERO g, ONE SECOND

Figure 3. Graphical representation of dependence of heat flux upon temperature difference between heater and boiling Freon 114 under one-sec zero and 1 g conditions.

TABLE 4: COMPARISON OF PHYSICAL PROPERTIES BETWEEN FREON 114 AND WATER

	Symbol	Units	Freon 114	Water
Molecular weight	M		170.93	18.02
Boiling point at 1 atm.		°C	3.55	100
		°F	38.39	212
Density Liquid at 30°C	ρ_L	g/ccm	1.440	1.000
Density Vapor saturated, at b.p.	ρ_ν	g/1	7.82	0.598
Heat of vaporization at b.p.	L	cal/g	32.78	538.7
Viscosity at 30°C	μ	centipoise	0.356	0.8007
Surface tension at 25°C	σ	dynes/cm	13	71.97

Shown are the 1-g nucleate and film boiling branches after equilibrium has been established, and the zero g boiling curve after one second of zero g. It was not possible to obtain data in the transition region between nucleate and film boiling at 1 g with an electrically heated wire, because the temperature cannot be controlled when a sudden change in film coefficient occurs.

Only one branch exists for zero g boiling, since nucleate boiling appears to be time dependent and is present only for a few milliseconds. It is felt that the zero g boiling curve, for longer zero g durations, will be shifted more into the 1-g equilibrium film boiling region of the diagram. Testing is currently under way to examine this zero g time dependence.

Equations and Nomenclature

$$(q/A)_{max} = \frac{\pi}{24} L\rho_v \left[\frac{\sigma g(\rho_L - \rho_v)}{\rho_v^2}\right]^{1/4} \left[\frac{\rho_L + \rho_v}{\rho_L}\right]^{1/2} \quad \text{(Zuber)}$$

$$(q/A)_{max} = KL\rho_v^{1/2}[g\sigma(\rho_L - \rho_v)]^{1/4} \quad \text{(Kutateladze-Borishanskii)}$$

$$K = 0.13 + 4N^{-0.4}$$

$$N = \frac{\rho_L \sigma^{3/2}}{\mu^2[g(\rho_L - \rho_v)]^{1/2}}$$

A area of surface of test section
E potential drop across test section
h coefficient of heat transfer from test section to boiling liquid
I electric current through test section
q heat transfer rate in test section
Δt temperature difference between heating wire and boiling liquid
ρ_L density of liquid
ρ_v density of vapor
L heat of vaporization
μ viscosity
σ surface tension
g gravitational acceleration
$(q/A)_{max}$ maximum heat flux for nucleate boiling

120 HANS F. STEINLE

References

1. Siegel, R. and Usiskin, C., "A Photographic Study of Boiling in the Absence of Gravity," ASME Paper 59-AV-37 and *Jour. Heat Transfer*, Vol. 81, Sec. C, No. 3, Aug., 1959.

2. Reynolds, W.C., "Behavior of Liquids in Free Fall," *Jour. Aero-Space Sciences*, Vol. 27, No. 12, Dec., 1959.

3. Benedikt, E. T., "Scale of Separation Phenomena in Liquids Under Conditions of Nearly Free Fall," *ARS Journal*, Vol. 29, No. 2, Feb., 1959.

4. Li, Ta, "Liquid Behavior in a Zero-g-Field," Convair-Astronautics Report No. AE60-0682.

5. Zuber, N., "On the Stability of Boiling Heat Transfer," *Trans. ASME*, Vol. 80, No. 3, April, 1958.

6. "Precision Resistance Thermometry," in *Temperature, Its Measurement and Control in Science and Industry*, N. Y., Reinhold Publ. Corp, 1941.

7. McAdams, W. H., Addoms, J. N., Rinaldo, P. M., and Day, A. S., "Heat Transfer from Single Horizontal Wires to Boiling Water," *Chemical Engineering Progress*, Vol. 44, No. 8, Aug., 1948.

ORBITAL FORCE FIELD
BOILING AND CONDENSING EXPERIMENT

ROBERT L. CUMMINGS, PAUL E. GREVSTAD, AND JOHN G. REITZ

New Devices Laboratories, TAPCO Group,
Thompson Ramo Wooldridge, Inc.

Throughout history, man's progress has been directly dependent upon the power available to him. This fact holds true as he enters the new environment of space. Capability for establishing useful scientific, commercial, and military satellites and for exploring the solar system will depend upon power for communication, control, guidance and propulsion of the satellite and space vehicles. To satisfy the power requirements of large space vehicles, it is certain that dynamic systems utilizing the Rankine cycle turbomachinery will be employed.

The Rankine-cycle method for converting thermal energy into electrical energy has long been the world's largest source of electrical energy. Nearly all large powerplants, burning a wide variety of fuels, use Rankine engines to convert this heat energy into a more usable form. When such a power-producing system is applied to space, however, the new environment and stringent mission requirements result in machinery differing considerably from the earthbound variety. The major environmental factors that dictate component design changes are the absence of an atmosphere (meaning that all waste heat must be rejected by radiation rather than by convection), and the complete absence of gravitational forces (necessitating a considerably modified approach to fluid mechanics considerations). Mission requirements dictate that the Rankine space power system provide electrical power reliably, for long durations, with minimum weight.

Figure 1 shows a block diagram of a Rankine space power system. Thermal energy, solar or nuclear, is used to vaporize the working fluid in the boiler; expanding this vapor through a turbine results in a portion of its thermal energy being converted to shaft energy which, in turn, is converted to electrical energy. The unusable thermal energy in the turbine exhaust vapor is removed in the

121

condenser, converting the working fluid back to liquid. This waste energy is radiated to space, and the condensed fluid flows to a condensate pump which returns it to the boiler. The components that may be affected by zero gravity are, primarily, the condenser, the boiler and the condensate pump. The dynamics of the mercury inventory in the complete loop must also be investigated.

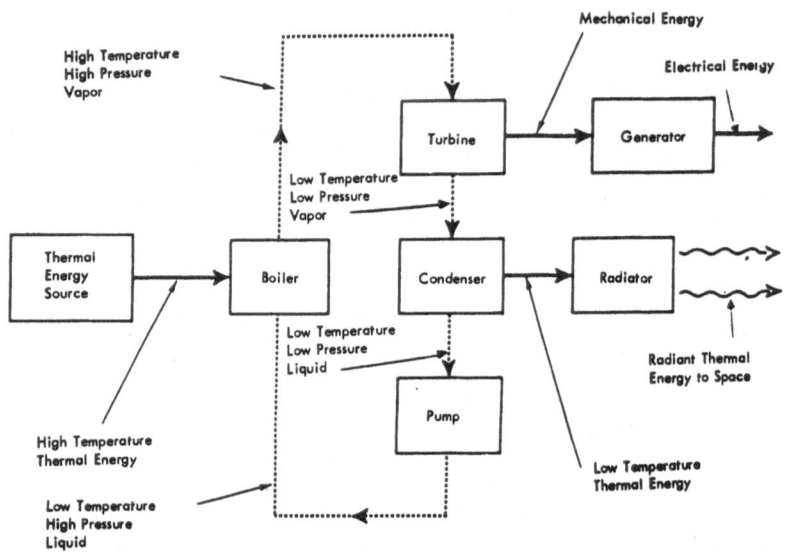

Figure 1. Block diagram of Rankine space power system.

Thompson Ramo Wooldridge, by virtue of its SNAP (Systems for Nuclear Auxiliary Power) programs, has been in the fortunate position of being able to participate in pioneering the use of the Rankine cycle in space. The AEC-sponsored SNAP I and SNAP II programs have been responsible for the development of analytical and experimental techniques, and hardware that have proved the feasibility and the potential of the Rankine system for this application. Active development has been underway at Thompson Ramo Wooldridge on the SNAP programs since early 1956, when feasibility testing began actually before Sputnik I. Performance goals have been achieved and the capability of this type of system for reliable, long-duration, unattended operation, has been recently very clearly demonstrated by the achievement of endurance demonstrations of 20 days for the SNAP II system and over three months continuous operation for the SNAP I system. The turbine-alternator package for SNAP I is shown in Fig. 2.

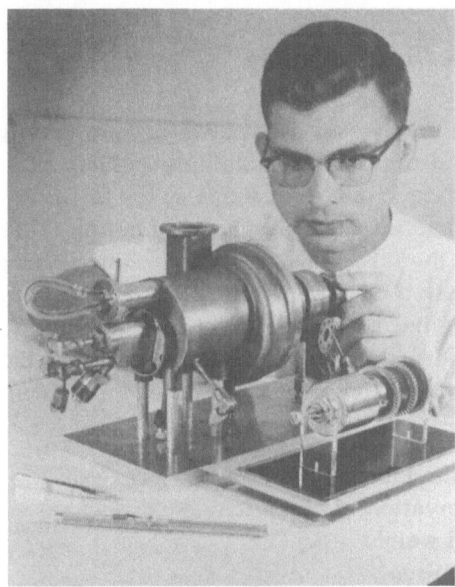

Figure 2. Turbine   alternator
package for SNAP 1.

It was to support the SNAP programs that in 1959 Thompson Ramo Wooldridge, under a company-sponsored program, began developing experimental apparatus to investigate mercury boiling and condensing in the zero gravity environment. All of Thompson Ramo Wooldridge's work to date has been dependent upon the ingenuity and cooperation of the Air Force personnel of the Flight Accessories Laboratory and the Aero Med Group at Wright Aeronautical Development Division, whose planes provide the zero gravity environment required. The first apparatus developed was flight-tested in WADD's C131B zero gravity aircraft.   Shortly after these pioneering tests, which will be described later, AEC support was obtained under the SNAP II program for the testing of more elaborate apparatus in zero gravity.   The AEC-sponsored effort has resulted in the design, fabrication, and ground testing of a forced convection mercury boiling, condensing and pump loop that

realistically simulates a complete mercury Rankine space power
system. This test rig, designed for use in the jet-powered KC-135
aircraft, will be given its first flight test in the very near future.
A continuing series of flight tests is planned to be made during
the summer and fall of 1960. (This paper was first given July 1,
1960.)

Methods for Conducting Zero Gravity Boiling and Condensing Experiments

To properly evaluate Rankine system phenomena in zero gravity
is a difficult task. The usual difficulties of obtaining reliable heat-
transfer and fluid-mechanics data with normal fluids in the laboratory
are multiplied several fold when attempts are made to obtain these
data with a liquid metal while the laboratory is in an aircraft gyrating
through the sky. Since zero gravity is the specific item of the space
environment that is required, the experimental apparatus must be
engineered to do the job under the requirements imposed by the
vehicle capable of providing this environment. Several design considerations are:

    a. Provide valid, repeatable data
    b. Permit control and observation of operation
    c. Be of minimum size and weight
    d. Have low power consumption
    e. Present no hazards for operating personnel or the vehicle

The merging of these requirements with the vehicles available for
providing zero gravity, has resulted in the two OFFBACE test
packages that will be described shortly.

There are also two principal characteristics that vehicles capable of providing zero gravity should possess to be most useful in
boiling and condensing experiments. The duration of zero gravity
should be as long as possible and of high "quality." While zero
gravity durations of a few seconds permit observation of fluid-flow
phenomena, a complete equilibrium is not obtained. Due to the
thermal masses of a Rankine system, the experiment ideally should
be in zero gravity for sufficient duration to assure thermal as well
as fluid dynamical equilibrium.

An OFFBACE test loop should also be provided with "high
quality" zero gravity in order to accurately obtain the data desired.
One of the areas most sensitive to any nonzero gravity disturbances

is the vapor-liquid interface in the condenser. In order that the behavior of this interface be representative of that which will occur
in a space power plant, it is essential for the vehicle not to tumble,
spin, or transmit vibrations or other accelerations to the test article.

There are, of course, very real difficulties in obtaining the desired long durations of pure zero gravity. Since an object in zero
gravity near the Earth's surface accelerates toward the Earth, increasing its velocity 32 ft/sec/sec, even short durations of zero
gravity result in high velocities. For example, one minute of zero
gravity results in a velocity change of about 1300 mph. Limiting
nongravitational forces acting upon the body to sufficiently small
values, while experiencing such velocities in the atmosphere, is
difficult indeed.

The two systems that Thompson Ramo Wooldridge has used to
implement its OFFBACE program have been designed specifically
for the zero gravity environment provided by Air Force transport-
type aircraft. The first system was evaluated in WADD's C-131 B
which provides approximately 12 sec of zero gravity. The second
system, just completing its shakedown tests, will be flown in the
WADD KC-135, capable of zero gravity durations in the order of
32 sec. Both of these test packages were so designed that engineers
could directly observe and operate them during flight. The KC-135
flight trajectory employed to obtain the zero gravity environment is
shown in Figure 4 (Schwartz) page 109.

For durations longer than the KC-135 can provide, it appears
advisable to utilize space available in ballistic missiles or in
orbital vehicles, with remote data collection and control consequently required. While higher performance type aircraft are capable
of providing up to about 1.5 min of zero gravity, they do not appear
to have the space available for a meaningful arrangement. Furthermore, the increase in duration is not enough to be really significant.
The larger aircraft have sufficient space to permit test apparatus to
be freely "floated" within the cabin, thereby, removing it from
"nonzero gravity" accelerations and vibrations.

The only method for obtaining extended durations of high quality
zero gravity is to utilize a vehicle capable of taking the test package outside the earth's atmosphere, i.e., sounding rockets, ballistic
missiles or satellites. These vehicles will provide not only zero
gravity, but the complete environment that a Rankine space power
system will experience. Limitations on payload weights and space

available for such a package, the severe launching environment, and greater complexity in data retrieval will make this form of testing more difficult than aircraft testing. Also, any spinning or tumbling would negate the pure zero gravity and would have to be controlled. However, these vehicles (especially large ballistic missiles and satellites, because of the possibility of low cost "rides") are the most attractive for later phases of OFFBACE type activities. Plans are being made for future programs to develop a suitable test package and telemetering instrumentation that will integrate with this type of zero gravity vehicle.

Mark I Test Program

The first apparatus developed to provide data on mercury condensing in zero gravity is shown in Figure 3. This apparatus is

Figure 3. Apparatus for observation of mercury condensing in zero gravity environment.

termed the Mark I OFFBACE Test Loop and was designed for testing in the zero gravity environment provided by the WADD C-131 B aircraft. Data from this loop are primarily in the form of motion pictures of the condensing action within a 12-in.-long glass condenser tube. The two 16-mm motion-picture cameras shown were usually used; one recorded the over-all view of the condenser, a timing clock, and a vertical accelerometer during the entire zero gravity maneuver. The high speed camera (FASTAX) recorded the condensing action in a short section of the tube.

Figure 4 shows the construction of the mercury loop and its enclosure. Mercury vapor is supplied by an electrically heated,

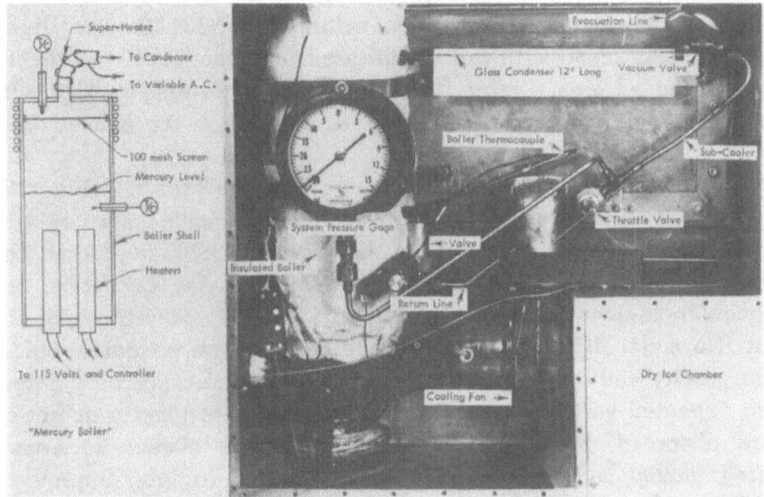

Figure 4. Mercury loop with enclosure for condensation experiments in zero gravity environment.

pool-type boiler fitted with a screen to prevent liquid from entering the condenser during zero and negative gravity. The vapor is condensed within an air-cooled Pyrex glass tube, and is returned to the boiler through the subcooler line. The valve in this return line is closed during the zero gravity maneuver in order that vapor flow to the condenser be maintained by the boiler's vapor capacity and the condenser cooling rate. Instrumentation, other than the motion-picture cameras, consists of thermocouples to indicate boiler, condenser inlet, and subcooler temperatures and a compound pressure gage to indicate system pressure level. The loop is housed within an enclosure to protect personnel from mercury vapor hazards in case of accident. Condenser cooling air is recirculated within this enclosure through a dry-ice chamber which serves as the heat sink.

The only major difficulty associated with testing this rig in zero gravity has been due to noncondensable gases within the system. Air trapped in the pressure gage and return line was compressed by hydrostatic pressure due to the mercury head prior to the zero gravity maneuver. During the tests, absence of gravity head allowed these gases to expand and move the vapor-liquid interface in the condenser tube toward the condenser inlet. Extra care in evacuating and sealing the system corrected this difficulty.

The majority of zero gravity testing of this apparatus has been with it rigidly mounted to a rack within the C-131 B. One flight, however, was made with the rig allowed to "free float" within the plane's padded area. During this flight, personnel restrained the rig from excessive motions and guided its descent to the padded floor at the conclusion of the maneuver.

The primary achievements of this experimental mercury condensing apparatus was to demonstrate that the condensate could be collected in a stable manner by viscous and pressure forces and that a definite and stable demarcation between the condenser and subcooler regions could be maintained for the conditions of the test. No major differences in the operation of the condenser under zero gravity and under ground environmental conditions were noted. Two expected variations between normal and zero gravity operation were observed. Since the condenser tube was always horizontal during ground and nonzero gravity portions of flights, droplets of condensate forming on the tube wall would flow diagonally down the wall as they were swept toward the interface. Most of the droplets, therefore, would be concentrated in the bottom of the tube. During zero gravity, the droplets were free from the orienting force of gravity and, therefore, flow was more evenly distributed over the flow area of the tube. This fact permits noncondensables to be more easily trapped and carried to the condensate pump during zero gravity operation. The vapor-liquid interface, slightly inclined during 1-g operation, was normal to the tube walls during zero gravity. The surface experienced small oscillations due to the impact of condensate droplets, but did not exhibit any large-scale unstable motion.

Mark II Test Program

In order quantitatively to investigate mercury boiling and condensing in components typical of Rankine space power systems, the

Mark II OFFBACE test loop has been developed. A schematic drawing of the Mark II loop is shown in Figure 5. It can be seen that this apparatus includes all of the Rankine elements except the turbine-alternator, which is the component least affected by zero gravity. The design of all components is such that they realistically represent concepts being employed in the SNAP II space power system.

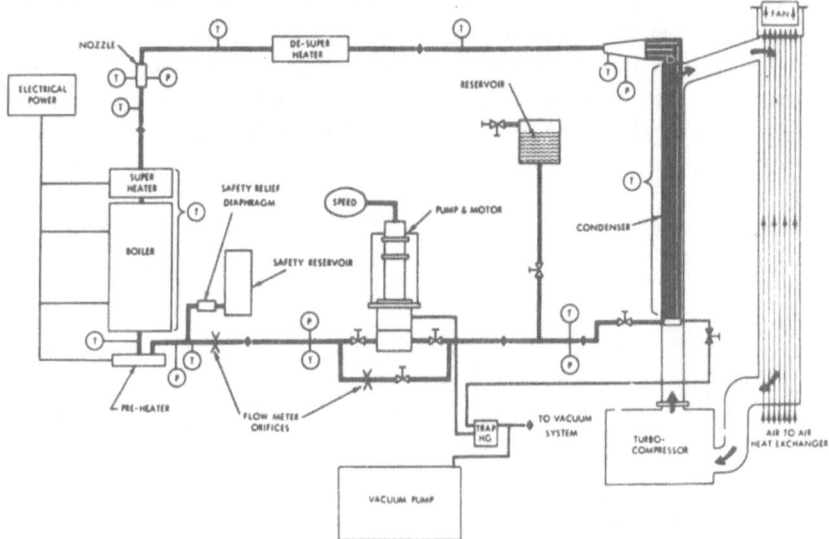

Figure 5. Schematic representation of Mark II OFFBACE.

The boiler is of the once-through type and employs direct electrical heating. A sonic nozzle and desuperheater simulate the turbine portion of the Rankine system, and also allow control of condenser inlet conditions. The design of the condenser is such that a wide variety of condenser designs can be tested with condensing tubes made of high temperature glass or metal. The mercury pump used is a SNAP II centrifugal type driven by a high speed electric motor.

Instrumentation includes strain gage transducers to measure inlet and discharge pressures of the condenser, pump and boiler; orifice flowmeters for the loop flow and bypass flow; pump speed; immersion and contact thermocouples for temperature measurements throughout the system; and strain-gage accelerometers in three axes. Motion-picture cameras are used for recording the condenser interface and over-all condensing action when glass condenser tubes are used.

The entire mercury loop is mounted within an enclosure as a precaution due to the potential hazards of mercury. Heat rejection from the condenser is accomplished by forced-air circulation over the condenser tubes, the air in turn being cooled in an air-to-air heat exchanger. The aircraft's cabin air is the final heat sink. Figure 6 is a photograph of the mercury loop of the Mark II rig taken during assembly. Also a part of the system is an instrumentation and control console remote from the mercury loop. These two units are connected by long, flexible cables so that the mercury loop can be operated while free floating within the aircraft.

Figure 6. Photograph of mercury loop of Mark II OFFBACE rig taken during assembly.

Major objectives of the Mark II OFFBACE testing are as follows:

a. Investigate condenser and boiler fluid mechanics and heat transfer in zero gravity conditions, and also in various other g conditions.

b. Investigate component interrelationships, particularly the condenser-centrifugal pump.

c. Investigate Rankine package problems such as pressure pickup in a high-temperature mercury vapor region, or any other apparatus element which could be affected by any g force, including zero, which could be encountered in aircraft, missile or satellite testing, for the ultimate purpose of successfully operating a mercury Rankine power system in outer space.

The Mark II system is presently being readied for installation into the KC-135 and flight testing will commence early in July.

Systems For Long-Duration Zero Gravity Tests

The design of systems to evaluate Rankine system phenomena in the longer-duration zero gravity state provided by missiles or satellites will be based on different concepts than those used for systems evaluated in the aircraft. Since such systems would be subjected to the complete space environment, it is certain that they will be developed and make significant contributions to the success of Rankine space power systems.

Significant departures from apparatus used for aircraft tests lie primarily in the areas of the condenser configuration, instrumentation and power supply. Since all heat must be rejected by radiation with this type of vehicle, the condenser configuration will have to provide the surface area to accomplish this. Instrumentation with telemetry must be provided to adequately evaluate the performance of the system. In addition to the type of system instrumentation discussed earlier, further measurements of the significant features of the space environment must be made. Recovery of the test package would be of no value except where data had been recorded within the package rather than telemetered, as might be the case with motion pictures of glass condenser tubes. The form of energy to supply the boiler might also be different for long-duration zero gravity tests than for aircraft tests. Direct electrical heating is utilized in aircraft because of its availability and convenience.

Batteries could be employed in missile or satellite tests, thereby requiring little modification to boiler heating schemes. However, weight limitations of missiles or satellites might require that the thermal energy source have a higher specific energy than batteries possess. For example, a satellite OFFBACE system to provide useful information for several days would not be feasible with batteries. For such systems and several others, the use of the heat of fusion of lithium hydride appears most promising for providing the required thermal energy, since its specific energy is about seven times as that of high energy batteries. For very extended test durations the use of a solar or nuclear energy will be required.

The OFFBACE system to be used for satellite and missile tests could be designed for use with a motor-driven pump, as in the Mark II test rig described. It appears very promising, however, to utilize the proven capabilities of the SNAP I or SNAP II system to provide the circulation and also to permit demonstration of the capabilities of a complete Rankine system in space.

Conclusions

The OFFBACE tests, using the Mark I test rig, have already given very valuable demonstrations of the operation of zero gravity condensers in tests in the WADD C-131 B aircraft. These tests have supplemented condenser testing done under the SNAP programs and have conclusively shown two important facts:

a. The condensate can be simply and reliably collected in a tubular condenser in zero gravity by utilization of viscous and pressure forces from the vapor.

b. A stable liquid-vapor interface can be maintained.

The forthcoming Mark II series of tests will provide further detailed component information during tests in the WADD KC-135. This rig will provide heat transfer and fluid dynamic results which will be fully typical of a complete Rankine space power system.

Further tests in satellite and ballistic vehicles are of major importance. The OFFBACE program using SNAP hardware may also be expected to make major contributions in this type of testing.

PART THREE

Zero Gravity Research in Bioastronautics

ZERO GRAVITY SIMULATION

RAPHAEL B. LEVINE

Lockheed Aircraft Corporation, Marietta, Georgia

During the past few years the United States Air Force has been conducting a series of experiments on the effects of weightlessness.* In these series the zero gravity condition is obtained by flying specially equipped aircraft in a precision maneuver (Keplerian trajectory) during which time the effects of nearly complete weightlessness are felt by the occupants of the plane. In fighter-type aircraft, the period of weightlessness may extend to 40 sec; however, in such a vehicle, the subject must be strapped into place. In larger aircraft, such as the KC-135, the period of weightlessness may be as long as 32 sec and the subjects can float freely. It is expected that the X-15 aircraft, now under test, will expose the pilot to as much as 4 min of weightlessness. The first Mercury astronaut experienced 5 min of weightlessness. The first reported Russian manned flight is said to have involved about 90 min of zero gravity.

Most investigators in the field of space physiology agree, however, that it would be dangerous to extrapolate from these experiments to conditions of long-term weightlessness. Some of the most serious of the effects on the human organism, which would involve moderate to total disablement in critical functions, may well appear only after from 20 min to several hours of exposure to null gravity. However, no way of obtaining such long periods of weightlessness was known, or until recently even postulated, short of actually putting a man in space and observing his reactions.

The fact that there is no way known, at least publicly, of suspending the law of gravity on the surface of the Earth has always been taken to mean that the effects of weightlessness lasting longer than a second or two cannot be studied in an earthbound laboratory. This paper proposes to show that such is no longer the case, and that it is possible to build an artificial environment in which the conditions of zero gravity are so well

*See the paper by E. L Brown in this volume.

135

simulated that many of the effects on living organisms of true long-term weightlessness may be studied before the date when orbiting laboratories are established.

It should perhaps be emphasized at the outset what such a "zero gravity simulator" is *not*. It is not a device to cancel the gravitational field, nor even to nullify its effects (as will be done in orbiting vehicles). It will not simulate weightlessness for *non-living* systems, except in very special cases where densities of rigid components approximate that of an immersion fluid or where viscosities of fluid components are very high.

What the simulator *is* designed to do is to produce nearly all of the psychological and many of the physiological effects of weightlessness in living organisms, and thus to make possible the study of such effects without the necessity of producing actual free-fall conditions for prolonged periods.

Effects of the Natural Zero Gravity State

On casual consideration, it is by no means obvious that any reasonably simple artificial environment can be successful in such a task. Nearly every body function is likely to be affected by the absence of weight. Some of the more important of these, from the standpoint of survival and performance in a zero gravity state, are listed in Table 1.

TABLE 1: SOME BODY FUNCTIONS AFFECTED BY WEIGHTLESSNESS

Alimentation	Manipulation
Cardiovascular function	Posture
Orientation	Vestibular function
Diurnal rhythms	Respiration
Rest and sleep	Muscle Tone
Locomotion	Psychological well being

The effects of weightlessness on these functions have been considered by many investigators,[1-23] and a good deal of information on this facet of space flight is now available.

Several effects emerge as critical, both in regard to the seriousness of the incapacity they might engender, and to their unpredictability in terms of present-day lack of knowledge. One of these is the matter of orientation in space. In the weightless state, it is possible to become completely disoriented within a very few seconds and the disorientation tends to persist during the remainder of the short periods of free-fall presently obtainable. It is certainly necessary to know whether such disorientation would eventually cease under continued zero gravity conditions. In addition, since orientation in free fall will depend nearly exclusively on the visual sense, it is not known whether sleep will bring on disorientation and whether artificial aids such as strap-in "bunks" or adhesive devices would overcome this loss of the visual sense.

Another critical effect is that of nausea or "spacesickness." The same receptors involved in sea- or air-sickness will be implicated if spacesickness develops, and it is already known that the latency of the ordinary forms is of the order of 20 min, even in the inexperienced traveler. It is necessary to know whether such a disability will develop in prolonged exposure to zero gravity, whether it persists, whether some individuals are immune to this effect, and whether tolerance can be developed in some individuals.

A third factor of much concern in space medicine is the psychological effect of such unusual factors as weightlessness, danger and isolation. What seems to be an exhilarating experience to many when experienced for a few seconds may seem quite different when prolonged for hours. Some of the effects of isolation and of danger have been studied, but not under the extremes to be encountered in early space travel, and certainly not in conjunction with weightlessness.

A second group of factors, not considered critical, nevertheless shows important gaps in current knowledge and should be studied further before manned space flight begins. One of the most useful of these is the work-rest cycle. Present meager information indicates that sleep under zero gravity conditions will be "fast;" i.e., that the equivalent of 8 hr of normal sleep might be accomplished in as little as 2 hr in space. This would have important connotations for the establishment of duty periods and of minimum crews for a fixed number of duties.

Other subcritical factors in this group include the matters of locomotion within a space vehicle in the absence of frictional forces and the manipulation of weightless (but not massless) objects by weightless operators. Depending on developments in space flight, locomotion outside a space vessel and to nearby objects in space may fall within this category or may be placed in the critical group.

A third group of factors is well-enough understood, or research has progressed to the point that they are not considered serious problems in space flight. Alimentation, cardiovascular function, the eyeball and its musculature, muscle tone, posture sense (knowledge of the relative arrangement of the different parts of the body), and respiration are not expected to suffer in space flight. Indeed, cardiovascular function, for instance, may even be improved in free fall.

The foregoing factors have been considered, thus far, in terms of their effects on the spacecrewman while he is in the weightless state. Perhaps even more serious are some of the longer-term effects of weightlessness, which may become apparent only, or especially, during re-entry periods or upon return to Earth-normal gravitational conditions. These include such factors as deterioration of muscles and bones, and reduced reserve capacity of the cardiovascular system during the protracted periods of reduced demand under zero gravity conditions.

Few, if any, of the severe effects mentioned can be studied during the short periods of free-fall obtainable in aircraft. Periods of real or well-simulated zero gravity lasting many minutes or hours, or even days, are necessary for such investigations.

Uses of a Zero Gravity Simulator

A simulator capable of providing an artificial zero gravity environment would be useful in many ways. Most immediate, perhaps would be its usefulness in selection of space crewmen on the basis of their inherent or acquired tolerance to the zero gravity state. A somewhat similar use would be as a training device for familiarization of prospective space crews with zero gravity conditions before actual take-off. Of great interest to many investigators would be the use of this artificial environment to study the physiology and psychology of weightlessness and other sensory deprivation conditions. Finally, results from such investigations as the foregoing should prove useful in the design of life-support systems for space vehicles, thus helping to reduce the stresses inherent in this new mode of transport.

In addition to the capacity for long-term studies of the zero gravity state, there are further advantages of the artificial environment as compared to methods of study already available. First, the simulator is safely on the ground, and convenient to complete laboratory facilities, even heavy and bulky ones. Second, the cost, especially when figured in terms of subject-minutes of weightlessness, is relatively very low. Third, there are no preliminary and subsequent high-g episodes, as in aircraft pulling out of dives, to mask the effects of weightlessness itself. Finally, the subject is not burdened with duties such as controlling the vehicle, and can concentrate on the directly applicable tasks assigned him.

Method of Zero Gravity Simulation

The Lockheed zero gravity simulator is expected to achieve effective artificial weightlessness by attacking two types of problems: the psychosensory problem, and the directly physiological problem of simulation.

The sensations associated with a gravitational environment arise from three major sets of cues, indicated in Table 2.

TABLE 2: GRAVITY-ORIENTATION CUES

Visual	Normal verticals and horizontals
Mechanical	Localized support, posture maintenance, visceral drag
Inner ear mechanisms	Otolith organ response

The first portion of the attack on the problem of simulation is to remove or nullify all of these cues. Visual nullification is accomplished simply by enclosing the subject, as in a tank, thereby depriving him of his view of gravitationally oriented horizontal and vertical lines such as seen in buildings, trees and the horizon.

Mechanoreceptor nullification is obtained through immersion of the subject in a fluid of specific gravity equal to his own (or the equally valid procedure of adjusting the subject's effective specific gravity to equal that of a convenient immersion fluid, such as water). This accomplishes a number of desirable results. In the first place, the localized pressure receptors in the skin are not selectively excited, as they normally are on the soles of the feet or the seat of the pants. The water support is equally distributed over the entire body. In the second place, the muscular effort of maintaining a posture against gravitational pull is reduced essentially to zero, especially if each freely articulated portion of the body is individually adjusted to neutral buoyancy by the addition of bands incorporating floats or weights as needed. Thus the cues to direction and magnitude of gravitational forces picked up by the muscle tension receptors are eliminated. Water immersion even tends to nullify a third set of mechanical cues from the stretch receptors in the connective tissue supporting soft tissues such as the abdominal contents. To the extent that these tissues have the same specific gravity as water (and this is very nearly the case) and to the extent that the abdominal wall is flexible and transmits hydrostatic pressure freely, the tendency of the belly wall to sag in the direction of the gravitational vertical is markedly reduced.

The fact that the combination of immersion and loss of visual cues is effective is borne out rather dramatically in many cases of skin-divers' disorientation. Knight[12] and Margaria[16] have studied these effects in detail, and have concluded that such a situation is a good first-order simulation for weightlessness.

However, the problem of gravitational cues from the otolith organs in the vestibules of the inner ears remains. These organs, located in the nonauditory labyrinths, as shown in Figure 1, are small and highly specialized for the detection of linear accelerations, including that of gravity. The semicircular canal organs, located in the same labyrinths, are likewise specialized, but for the detection of angular accelerations of the head.

Neither of these sets of organs is affected by water immersion

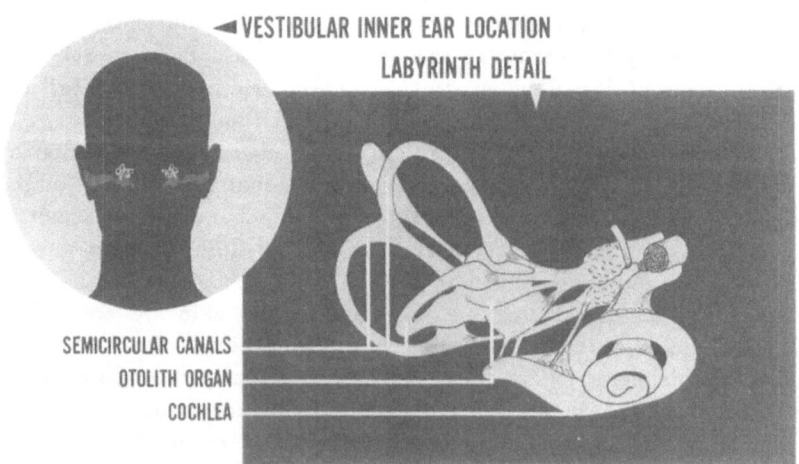

Figure 1. Vestibular apparatus.

of the external body, but the otolith would strongly feel any actual nullification of the pervading gravitational field. Further, as Margaria[16] has shown, skin-divers' disorientation does not persist in experienced subjects, and in man the otolith cues are very effective (though of low precision) when they are the only cues available. Thus, an effective simulator must include a way to nullify the otolithic cues.

Three investigators apparently arrived at the basic idea for such nullification independently, early in 1958. They were H. J. Muller, the Nobel prize-winning geneticist at Indiana University,[32] Otto Schueller, an engineer at the Wright Air Development Division,[33] and Ralph W. Stone, of the NASA Space Task Group at Langley Field.[34] In addition, the last of these developed a qualitative biophysical justification for the concept. Basically, all three visualized the human subject as suspended in a tank of water, with the tank, water and subject all rotating at the same speed about some horizontal axis. Figure 2 illustrates the Muller and Schueller concept, as modified in the Lockheed Human Factors Research Laboratory. The Stone approach uses a vertical tank with the subject

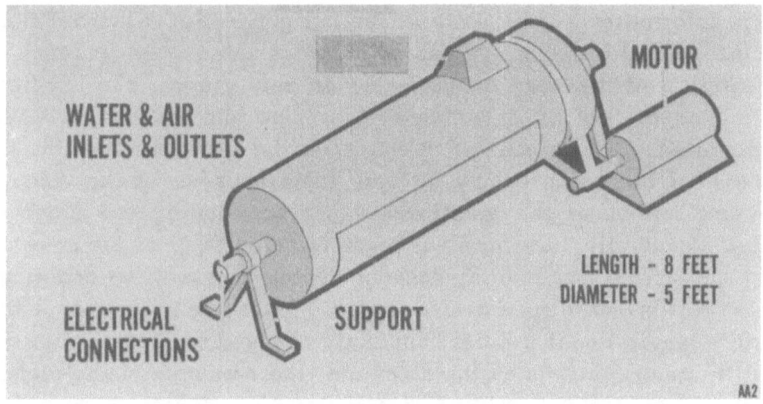

Figure 2. Zero gravity simulator, exterior configuration.

seated inside, his head being located at the center of the tank's height. Rotation is about a horizontal axis through the subject's ears.

A biophysical analysis of the otolith structure, shown in Figure 3, indicates why rotation should be successful in nullifying otolith

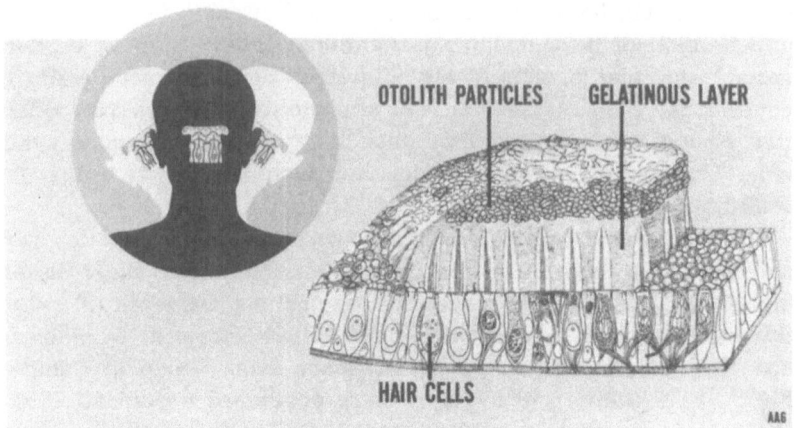

Figure 3. Otolith organ.

response.   Linear acceleration or field forces acting on the heavy
otolith particles cause them to move with respect to the hair cells,
thus bending the brush of hairs and distorting the cells.   Nerve
fibers surrounding the cells are activated by this distortion and
carry information on the state of the net gravitational-inertial field
to the central nervous system, where it is interpreted in terms of
orientation of the head to the field.   In zero gravity, the net field
acting on the otolith particles will be zero and the hairs will be
undistorted.   The simulator operates on the principle that the response of the organ cannot be very rapid because of the viscous
damping effect of the gelatinous layer surrounding the brush of
hairs. Physically, therefore, this organ is an example of the mechanical engineer's mass-spring-dashpot system and may be described
by a well-known differential equation.   When the direction of the
earth's gravitational field is constantly changed with respect to the
otolith organ, as it is in the simulator, the response of the otolith
organ is analogous to that of zero-center d-c voltmeter upon which
is impressed an a-c voltage.   The needle will follow slow alterna-
tions, but as the frequency rises the needle follows with smaller and
smaller fluctuations about its zero point (and with greater and
greater phase lag). By raising the frequency sufficiently, the needle
can be kept as near the zero position as desired.   Similarly, by rotating the subject sufficiently fast, his otoliths are kept as near the
zero deflection position as desired.   Thus, under these circum-

stances, the organ is kept essentially in its physiological zero gravity state, that is, with zero deflection of the hairs.

Revolving of the subject about a horizontal axis through the otoliths should therefore nullify the otolith response, provided that the rotation is sufficiently fast. In turn, the rate of rotation required depends on the damping coefficient, or time constant of response, of the otoliths. Only a small amount of data bearing on this matter was available until very recently. The work of Graybiel and Brown,[28] Gerathewohl and Stallings,[7] and Levine[30] giving information on otolithic time constants indicated that a rotation speed of about 10 rpm would be sufficient to damp out all but about 5 per cent of the otolith response to the alternating force field. However, later work by Gray[27] indicated that a sufficient speed would be above 30 rpm. Both sets of results indicated, however, that the proposed nullification technique would be effective.

Thus it appears, at least theoretically, that the attack on the psychosensory aspect of the simulation problem will be successful. A considerable degree of success is also expected for the physiological aspect of simulation. In the first place, if disorientation, vertigo or nausea are to be encountered in true space flight, the fact that the otolith organs will be in the same condition in the simulated as in the actual zero gravity situation leads to the expectation that these untoward reactions would occur in the simulator also. In the second place, as has been mentioned, the effects of immersion on the subject's musculature will be very much like those resulting from true free fall. A prolonged immersion experiment performed by Graveline[26] has shown that the skeletal muscles do indeed atrophy to an alarming extent in only a week. In addition, an even less accurately predicted result of this experiment was that the bones also began to deteriorate as indicated by the excretion of large quantities of calcium and phosphorus. The subject's ability to withstand the stresses of forces 1 g and greater in magnitude was also seriously decreased in this time. In the third place, water immersion will, to a very large extent, nullify the development of hemostatic pressures normally arising in a gravitational field. In free fall, of course, there will be no hemostatic pressure differences within the circulatory system. In water immersion, the external water pressure will act to counter the internal blood pressure to the extent that these two fluids have the same density. Finally, the interesting conditions of sleep needs and of work and rest cycle capabilities should be amenable to study in this simulator. It is postulated, for instance, that under such conditions subjects will need as little as two hours of sleep per day in order to be capable of performing effectively during the other 22 hours. The foregoing and other considerations make it appear very likely that some of the important physiological effects of extended free fall will be duplicated to a considerable extent in the zero gravity simulator.

Experimental Test of Otolith Nullification

The complete feasibility of this approach to the study of long-term weightlessness will not be known fully until preliminary experiments using human subjects are run in the simulator. However, some studies have been made on actual, functioning otolith organs, to test the otolith nullification hypothesis. The experimental animal was the fish, whose otolith organs are qualitatively very similar to those of human beings, and the apparatus was a scale model of the zero gravity simulator. The experimental set-up is shown in Figure 4.

Figure 4. Fish in scale-model zero gravity simulator.

These experiments have been reported elsewhere[31] and a summary of the procedure and findings follows.

Eight five-inch minnows were used in this series of experiments. The experimental procedure is illustrated in Figure 5. The rotational speed of the tank is increased until the fish successively

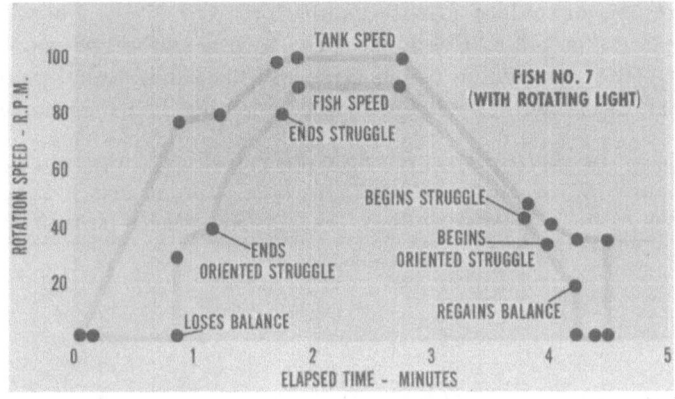

Figure 5. Typical otolith nullification run.

loses its equilibrium, its gravitational orientation, and finally ceases to struggle. The fish is kept at this high rotational speed for periods of a minute or more during which, it should be emphasized, the relative motion of fish and tank is very small. Insofar as swimming movements are concerned, this situation is the equivalent of being in still water; however, no such movements are observed. Since each fish observed struggles strongly against tilting away from the gravitational vertical, it is assumed that cessation of struggling means that the fish no longer possesses any cues to that vertical. This assumption is strengthened by the fact that in the second portion of the experimental run, involving decreasing rotational speed, the fish resumes the struggle at lower speeds. During the first portion of this decrease in speed, the fish continues relaxed, and its speed of rotation also decreases proportionately. When the speed has decreased considerably below that originally necessary to disorient the fish, the animal exhibits the following behavior successively: unorganized struggling movements (this is assumed to be the threshold for gravitational sensation); struggling movements which are oriented to the gravitational vertical; and finally swimming motions which successively achieve and maintain normal posture in spite of continued rotation of the water. The run ends shortly thereafter when the tank rotation is stopped.

In the experiments reported here, the only illumination was provided by a pair of small lights located at the periphery of the tank and rotated with the tank (see Figure 4). This was done to deprive the animal of gravitationally oriented visual cues. Each fish was subjected to five runs of the type described above. Table 3 shows the mean value of the threshold for gravitational sensation for each fish, together with the average deviation from the mean value.

TABLE 3: ROTATION SPEED FOR OTOLITH NULLIFICATION IN FISH*

Fish No.	Mean value	Average deviation from mean
1	40	6
2	39	6
3	40	2
4	63	3
5	49	3
6	77	5
7	45	7
8	57	1
Average	50	4

*Descending threshold in rpm for orientation struggle.

These data seem to confirm the hypothesis of otolith nullification by spinning. It is not possible, of course, to apply the numerical results of these experiments to the human subject, especially since there are large individual differences among the fish tested. However, the high degree of structural similarity between fish and human otolith organs seems to justify the qualitative extrapolation of these results to the human case. It may be assumed, with some confidence, that otolith nullification in the human subject will be attained at some reasonable rotational speed.

Construction of the Zero Gravity Simulator

Basic biophysical and human engineering considerations resulted in a simulator preliminary design* shown in Figure 6. The dimensions were specified as approximately 5 ft in diameter by 9 ft long.

*Simulator design and construction sponsored by Lockheed Aircraft Corporation, Georgia Division.

Figure 6. Zero gravity simulator, preliminary design.

The tank would therefore hold about 1300 gal or 11,000 lb of water. It would be capable of rotation speeds up to 80 rpm. The subject would be provided with respiration equipment, and with safety gear for rapid removal from the tank in case of emergency.

To study the hydrodynamics of such a tank, the scale model previously shown in Figure 4 was constructed. The scale factor is 1 to 7.5, and the model showed some of the effects on the movement of the enclosed fluid of such factors as smooth walls versus baffles, of subject balance and buoyancy, and of speed of rotation. The data derived from this study are in terms of Reynolds numbers applicable to the full-sized tank.

Some of the other practical considerations in the design and construction of an artificial environment of this sort are indicated in Figure 7.   The tank is operated from a remote-control console, which also includes instrumentation for monitoring the mechanical functioning of the simulator and the physiological state of the subject. A closed-circuit television system shows the subject's facial expressions at all times. The subject is provided with underwater breathing equipment which receives and exhausts air from an external source and sink. Neutral buoyancy is achieved through use of flotation bands on upper and lower arms, upper and lower legs, head, hips and chest. Buoyant or dense material is used in the bands, as appropriate to the portion of the body being adjusted. The subject is shown wearing a skintight, full-coverage suit for protection of the skin from the effects of long-term water immersion. For short experiments, no special protection will be required. The subject is represented as holding himself in a fixed position with one hand and as operating a tracking task with his other hand. Before him is a task performance panel, with reporting controls, for the study of cognitive and psychomotor performance during weightlessness. Besides television monitoring, the subject is in voice communication with the experimental crew.

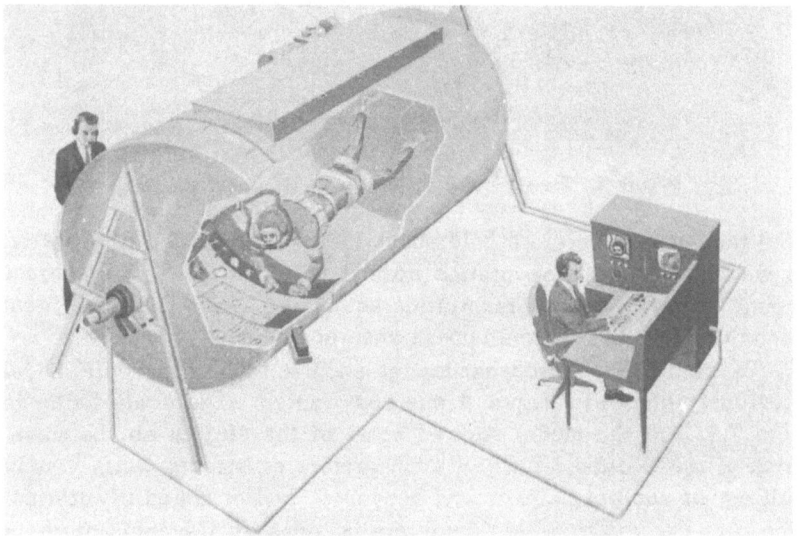

Figure 7. Zero gravity simulator design concept.

Also shown are six loose cords attached to the subject's harness at shoulders, hips and legs. These cords terminate on a six-drum winch built into the entry hatch cover. Another set of cords, not shown in Figure 7, continues to the opposite side of the tank and terminates in a spring-driven reel. This reel acts to pull the subject into the tank when the winch unwinds. The reel may be locked at any position, and the winch can then pay out slack to release the subject from restraint, as shown in Figure 7, or the winch can wind in all slack, to restrain the subject firmly in place. Both the winch and the reel lock are controlled by the remote operator, so that positioning of the subject can be accomplished whether the tank is rotating or motionless.

The final design has followed the original design concept rather closely, as may be seen by comparison of Figures 7 and 8. The latter figure shows the simulator in its laboratory setting, in a fairly advanced stage of construction. Visible in this photograph are one of three observation portholes, half-silvered so that the subject may be observed without being able to see outside; the entry hatch, hinged at the near side (see also Figure 6); the hatch locking mechanism; the subject winch motor and gear box, set into a recess in the hatch cover; the band of unpainted metal against which an air-brake shoe bears; one of three emergency dump valves (at the far and lower end of the tank); several of the 12 small portholes, each surmounted with a lamp and housing, through which sequencing visual cues may be given to the subject; the A-frame bearing supports for the tank; the driving belts and pulleys, the smaller of

Figure 8. Zero gravity simulator in laboratory.

which is mounted on the output shaft of a reduction gear box, just visible in the picture; and a portion of the plumbing arrangement which conducts temperature-selected water into the tank through a standpipe, which in turn allows the subject to make his necessary breathing movements by passing an equivalent volume of water into and out of the tank to the standpipe. Not visible is the 10-hp electric motor which drives the reduction gear through a magnetic clutch.

In an adjacent room a fixed, open tank is located. This tank is used for checkout of the subject's flotation gear, his respiration and communication apparatus, and the physiological monitoring sensors. The subject will be adjusted to neutral buoyancy and become familiar with his gear at the same time, while in this tank.

The matter of subject safety has constituted one of the major design considerations of the zero gravity simulator. This equipment, like the widely used human centrifuges, is potentially of great danger to subject and observers if improperly handled or poorly maintained. Extensive precautions have been taken to provide basic safety measures, back-up safety measures, and even tertiary back-up safety measures. Training of operating crews will be a major task to assure both effective operation and safety for subject and bystanders.

In an emergency, whether caused by subject illness or panic or by the failure of some component of the simulator or laboratory facility, the subject may be removed from the equipment in from 15 to 30 sec. This is accomplished in a sequence of four steps: first, the spring reel is unlocked, removing slack from restraints; second, the winch is activated to pull the subject over to the inside of the hatch cover and to snub him firmly to that surface; third, the tank and subject are stopped by means of the air brake (although the water continues rotating); and fourth; the hatch, with the subject in it, is opened to the position shown in Figure 6.

Conclusions

If fully successful, this artificial environment will permit a large number of very significant studies to be made. Native tolerance to the zero gravity condition, or its development, and a concomitant ability to orient oneself by visual cues alone, will be among the first and most important reactions to be studied. Damping coefficients and thresholds of the vestibular mechanisms to small lineal and angular accelerations may also be studied with this simulator.

Lockheed's Human Factors Laboratory is especially interested in human performance under conditions of stress and fatigue and the elaborate facilities already developed for such studies[29] would be used with the zero gravity simulator. Work and rest cycles, another Lockheed study area,[24,25] and sleep requirements, which are expected to be significantly different under weightless conditions, form another important group of studies. The effects of various antiemetic, ataractic and soporific drugs on all these reactions may be examined advantageously with this environment. The effects of weightlessness on several other physiological processes, such as digestion, cardiovascular and higher integrative functions also offer challenging problems which may be resolved with the zero gravity simulator.

It was expected that preliminary checkouts of operating and safety equipment would be completed during June 1961, and that the first manned runs would be made shortly thereafter. Following this, a series of feasibility studies is planned, to examine conditions for otolith nullification; to test the capability of the simulator to produce prolonged weightlessness conditions; to study subject zero gravity disorientation effects; to study the effectiveness of various visual, tactile and otolithic orientation cues; and to study vestibular thresholds. If the simulator performs as expected, it should be the forerunner of many such installations. The effects of the prolonged subgravity state will, in that event, become an important earthside study, much as the study of hypergravity states has been, through the use of centrifuges.

References

1. H. J. A. von Beckh, "Experiments with Animals and Human Subjects under Sub- and Zero- Gravity Conditions during the Dive and Parabolic Flight," *J. Aviat. Med.* 25, pp. 235-241, 1954. Later published as "La Gravitacion Cero," pp. 99-110 in "Fisiologia del Vuelo," Alfa, Buenos Aires, 1955.

2. Edward L. Brown, and M. R. Rockway, "Research on Human Performance during Zero Gravity," *Human Control Dynamics in Air and Space Craft: Proceedings Second Annual International Air Safety Seminar,* Nov., 1958, N.Y., Flight Safety Foundation, Inc.

3. Arthur C. Clarke, *The Exploration of Space.* N.Y., Harper & Bros., 1951.

4. G.T. Crampton, "Vestibular Physiology and Related Parameters in Orbital Flight." Symposium on Possible Uses of Earth

Satellites in Life Sciences Experiments, AIBS, Nat. Acad. Sci., NSF. Wash. D.C., May, 1958.

5. S.J. Gerathewohl, "Physics and Psychophysics of Weightlessness—Visual Perceptions,"*J. Aviat. Med.*, 23, pp. 373-395, 1952.

6. S.J. Gerathewohl, H. Strughold, and H.D. Stallings, "Sensomotor Performance during Weightlessness; Eye-Hand Coordination," *J. Aviat. Med.*, 28, pp. 7-12, 1957.

7. S.J. Gerathewohl and H.D. Stallings, "The Labyrinthine Posture (Righting) Reflex in the Cat During Weightlessness," *J. Aviat. Med.* 28, pp. 345-355, 1957.

8. Heinz Haber, "The Human Body in Space" *Scientific American*, 184, pp. 16-19, Jan., 1951.

9. Heinz Haber and S.J. Gerathewohl, "Physics and Psychophysics of Weightlessness," *J. Aviat. Med.*, 22, pp. 180-189, 1951.

10. J.P. Henry, E.R. Ballinger, P.J. Maher, and D.G. Simons, "Animal Studies of the Subgravity State During Rocket Flight," *J. Aviat. Med.*, 23, pp. 421-432, 1952.

11. F.A. Hitchcock, "Some Considerations in Regard to the Physiology of Space Flight," *Astronautica Acta*, 2, pp. 20-24, 1956.

12. Leon A. Knight, "An Approach to the Physiologic Simulation of the Null-Gravity State," *J. Aviat. Med.*, 29, pp. 283-286, 1958.

13. A.G. Kousnetzov, "Some Results of Biological Experiments in Rockets and Sputnik II," *J. Aviat. Med.*, 29, pp. 781-784, 1958.

14. Edwin Z. Levy, George E. Ruff, and Victor H. Thaler, "Studies in Human Isolation," *J. of the Am. Med. Assoc.*, 169, pp. 236-239, 1959.

15. John P. Marbarger, *Space Medicine: The Human Factor in Flights Beyond the Earth*, Urbana, Univ. of Ill. Press, 1951.

16. Rodolfo Margaria, "Wide Range Investigations of Acceleration in Man and Animals," *J. Aviat. Med.*, 29, pp. 855-871, 1958.

17. George E. Ruff, "Isolation," *Astronautics*, Vol. 2, pp. 22-24, Feb., 1959.

18. D.G. Simons, "Review of Biological Effects of Subgravity and Weightlessness," *Jet Propulsion*, 25, pp. 209-211, 1955.

19. Alan E. Slater, "Sensory Perceptions of the Weightless Condition," *Realities of Space Travel*, ed. L. J. Carter. Selected papers of the British Interplanetary Society, London, Putnam, 1957.

20. Hubertus Strughold, "Medical Problems Involved in Orbital Space Flight," *Jet Propulsion*, 26, pp. 745-756, 1956.

21. Hubertus Strughold, "Mechanoreceptors, Gravireceptors," *J. of Astronautics*, 4, pp. 61-63, Winter, 1957.

22. S. Gordon Vaeth, "Training for Space," *Astronautics*, 1, pp. 1-6, 30-32, 1954.

23. J.E. Ward, W.R. Hawkins, and H. Stallings, "Physiological Response to Subgravity States: II. Mechanics of Nourishment and Deglutition of Solids and Liquids," USAF-SAM Research Report, *J. Aviat. Med.* 30, pp. 151-154, 1959.

24. O.S. Adams, R.B. Levine, and W.D. Chiles, "Research to Investigate Factors Affecting Multiple-Task Psychomotor Performance," WADC Tech. Rep. 59-120, March, 1959.

25. O.S. Adams and W.D. Chiles, "Human Performance as a Function of the Work-Rest Cycle," WADD Tech. Rep. 60-248, March, 1960.

26. D.E. Graveline, (reported by D. Zylstra), "Astronauts Will Require Less Sleep," *Missiles and Rockets*, pp. 33-34, Feb. 29, 1960.

27. R.F. Gray, "Preliminary Study of Damping of the Otolith Organ System by Epicyclic Rotation," Naval Air Development Center MA-5919, Task MR 005. 13-6002.1, Report No. 10., Dec., 1959.

28. A. Graybiel and R.H. Brown, "The Delay in Visual Reorientation Following Exposure to a Change in Direction of Resultant Force on a Human Centrifuge." Rep. No. 1 USN School Aviat. Med. Res. and Tulane Univ. Joint Rep. No. 3, 1949.

29. J.A. Kraft, "Measurement of Stress and Fatigue in Flight Crews During Confinement." *Aerospace Med.*, 30, p. 424, June, 1959.

30. Raphael B. Levine, "Oculogravic Time-Constant," unpublished data, May, 1959.

31. Raphael B. Levine, "Null-Gravity Simulation," Operations Research Div. Rept. ORD 232, Marietta, Ga., Lockheed Aircraft Corporation, 1960, presented at 31st Annual Meeting of the Aerospace Medical Assn., Miami, May, 1960.

32. H.J. Muller, "Aproximation to a Gravity-Free Situation for the Human Organism at Moderate Expense," *Science*, 128, p. 772, 1958.

33. O. Schueller, "Space Simulators," paper in *Vistas in Astronautics*, Vol. II, 1958.

34. R.W. Stone, personal communication, Sept., 1959.

CURRENT WADD WEIGHTLESS RESEARCH

J. C. SIMONS*

*Capt., USAF, Wright Air Development Division,
Wright Patterson Air Force Base, Dayton, Ohio*

Let us first examine the scope of the ARDC zero g research program. As of May 1960, about 140 research programs in weightlessness were being conducted by ARDC (WADD), while 19 outside programs were being sponsored. The goals of the ARDC program for 1958-1965 are as follows:

(a) Determination of problem areas
(b) Cooperation with industry on these problem areas
(c) Determination of adequacy of various facilities available for use for zero g simulation
(d) Validation of simulator research in aircraft
(e) Instrumentation for orbital flight
(f) Development of recording techniques
(g) Integration of results

Ground and Airborne Facilities

The available facilities fall into two categories—the Flight Facilities and the Ground Center. In the future, a space flight facility is being added through access to the Scout, Atlas, and Mercury Programs. Presently available ground facilities do not, of course, achieve true simulation of zero gravity, but only of some of its effects. In this category we might mention the frictionless platform or scooter, and the submersion device (see paper by H. Levine in this volume). Returning to the flight facilities, the duration through which zero gravity situation can be maintained in the various planes used for the purpose are given in Table 1.

Much longer durations can be obtained with missiles (X-15, Atlas). Of the planes listed in Table 1, the KC-135A is probably the one figuring most prominently in the program. Electrical power available on board this plane is adequate for almost any need. The power output from one of the jet engines on the aircraft can be

*Paper presented at the symposium by M. S. Gardner, Capt., USAF.

utilized for energizing the experimental devices on board. A complete outfit for photographic (black and white) purposes, a closed-circuit television, and a 24-channel oscillograph are available. The aircraft is completely padded, so that protection for more delicate experimental equipment is provided.

Approximately 20 to 30 runs can be carried out during a 3-hr flight. Usually two parabolas in a row are flown, unless it is absolutely necessary to fly one at a time.

Any amount of acceleration between nearly zero and 2.5 g can be obtained. The length of period through which a zero g environment can be simulated is a function of the speed of entry (longer durations of zero g situations can be obtained by increasing the Mach number at entry). The optimal speed of entry for the F-100 aircraft was found to occur at Mach number .94; under those conditions a longer time at altitude, and more frequent passes can be obtained than if the entry has been supersonic. All the flights are conducted in a radar controlled area.

TABLE I: Duration of Weightlessness Obtainable With Various Types of Aeroplanes

Type of Plane	Duration of Weightlessness, Sec
31-B	12-15
K135-A	31 (approx)
TF-100	over 60
F14-B	90

Zero Gravity Research Information Center

In order to coordinate its cooperation with industry, WADD has been instructed by ARDC to set up a zero gravity research information center. This center will function as a central clearing house for USAF and NASA zero g research; it will conduct reviews of proposed, current and completed studies; it will advise participants in the program regarding the types and characteristics of facilities available, and attempt to prevent duplication of effort. The Center will also publish a handbook with periodic revisions to be distributed throughout NASA and ARDC as well as to participants in the work of the Center. This handbook will actually not appear as a separate entity, but as part of a more comprehensive publication called "New Regimes of Flight." Great reliance is placed for the success of the program, upon the initiative and effort of the participants.

HUMAN PERFORMANCE AND BEHAVIOR
DURING ZERO GRAVITY

EDWARD L. BROWN

Major, USAF, Air Force Ballistic Missile Division (ARDC)

During his recent assignment to the Aerospace Medical Laboratory at the Wright Air Development Division, the author initiated
and supervised a research program aimed at determining human
and system performance during zero g. These studies, which are
still continuing, took place, for the first two years, in a C-131B
transport-type airplane while it was flying a Keplerian trajectory.*
This airplane can produce about 15 sec of zero g. The research
program has recently been accelerated by the addition of a KC-
135, which will produce about 30 sec of zero g. This paper will
summarize a major portion of the work conducted in the C-131B.

The research areas that were investigated during zero g include
the following:

(a) Human performance on motor and mental tasks
    1. Normal flying performance and procedures during zero g
    2. Performance on actual emergencies during zero g
    3. Human performance on experimental tasks
(b) Locomotion of individual humans inside large space vehicles
(c) Locomotion of individual humans outside space vehicles
(d) Human perceptive orientation during zero g
(e) Behavior of liquids during zero g
(f) Fluid transfer problems during zero g
(g) Heat transfer problems during zero g

Human Performance on Motor and Mental Tasks

The prime point to keep in mind when considering motor and
mental tasks during zero g is that more than two thousand zero g
trajectories, each lasting from 10 to 15 sec were flown. While it
takes a few trials for a pilot to learn to fly a good zero g trajectory,
the task does not appear to be much different from any other precision instrument flying task, such as an instrument low approach.

Normal Flying Performance and Procedures

The degree of accuracy in flying the trajectory that can be
achieved by an experienced pilot is better than ± .05 g, using a

*Typical maneuvers of this type are illustrated elsewhere in
this volume.

specially modified mechanical instrument. Preliminary results with a specially designed indicator employing a spherical cork floating in a plastic container of water indicate that we can expect to come even closer to absolute zero g in the future.

The procedure that has been found to be satisfactory for flying zero g divides the task among the pilot, copilot and flight engineer. The pilot, monitoring the attitude indicator, zero g indicator and airspeed, places the plane in a 10-deg dive, holding this until a speed of 250 knots is reached. At this time he pulls the plane into a 35-deg climb angle, holding about 2.5 g's. As soon as 35 deg is reached, he pushes forward on the control column until the g indicator indicates zero. The zero is held until a dive angle of 35 deg is reached, at which time a 2 to 2.5 g pull-out is initiated.

The copilot's primary task is to control longitudinal g, which he does by controlling the power settings. When the plane is pulling up to the 35-deg climb angle, he reduces the throttle setting to about 18 in. of manifold pressure. Experience has indicated this setting is just about right for reducing the longitudinal accelerations during the maneuver to zero g. When he completes this setting, he then assists the flight engineer in monitoring the engine instruments, concentrating primarily on the oil pressure indicators. If the oil pressure drops below the red line, he, or the flight engineer, whoever notices it first, says "Oil pressure." This is the signal to the pilot that his oil pressure is failing and he must initiate an immediate control action that will produce a positive g acceleration on the airplane.

The flight engineer's prime job is to watch the engine instruments during the complete maneuver, calling out any incipient malfunctions to the pilot.

These procedures have worked satisfactorily for the last 1800 zero g trajectories. With tight safety belts, the pilot and copilot can perform their duties at zero g as easily and as accurately as during normal g.

Performance on Actual Emergencies During Zero G

Actual emergency situations have occurred on two occasions. Both of these involved overspeeding propellers, an item that is potentially very hazardous. The first time this occurred it came as a complete surprise to the two pilots, one of whom was on his first zero g flight and had acted as copilot for three trajectories. On the fourth trajectory, he acted as pilot and overcontrolled, unknowingly putting the airplane into a trajectory with a very slight negative g component. About two thirds through the trajectory both propellers began to overspeed. The pilots, while flying in essentially a zero g condition, had about two sec in which to analyze the situation and perform the correct action of feathering the propellers, which they did. One of the propellers, however, due to the complete loss of oil pressure in its system, continued to overspeed, even though the pilot had taken correct action. It was later determined that close monitoring of the oil pressure provided more lead time for initiating the propeller feathering, and on a subsequent flight in which a propeller began to overspeed it was brought under control before it approached the danger limits. These incidents provide us with some evidence that people can function correctly in an emergency situation during zero g. With experience and training in specific emergency situations, they can anticipate trouble and take corrective action before the emergency becomes serious, just as they can during normal g conditions.

Human Performance on Experimental Tasks

Several experiments on motor tasks have been performed during zero g. These have involved rapid movements of levers and switches, performed with tight safety belts and with loose safety belts. In general, with a tight safety belt, a man can perform as accurately and as quickly during zero g as during normal g. The major effect of a loose safety belt is to cause him to use one hand to hold onto any convenient solid apparatus to retain his position while he performs the task with the other hand. Relatively little force is required to operate the switches and levers involved in this experiment. It is expected that as the force for moving levers goes up, the requirement for a tight safety belt increases. The motion pictures taken of subjects performing during zero g also indicate the desirability for having a stirrup-type arrangement for the toes to prevent the feet from rising during the performance of motor tasks. While it has not been investigated, it appears that something that would assist in holding the feet solidly in place and about 12-14 in. apart, as well as a tight safety belt, would be of assistance during motor tasks requiring large forces to be applied through the arms and hands.

A task approximating a blind positioning-type task has been investigated. In this task the subject was sitting at the panel with his safety belt fastened. A piece of paper with a target on it was placed on the wall on his left, about opposite his hip and about 10 in. away from his body. The task was to alternate between actuating a switch on the panel and making a mark on the paper, as close to the target as possible without looking at the target. The eyes were kept fixed on the panel. A pencil was held in the left hand for making the mark and the left hand was also used for the switch actuation. So far, only one subject has performed the experiment. He practiced the task for about 30 sec during normal g, watching his hand as he attempted to mark the target. He then performed that task for about 10 sec, not watching his hand when it was in the target area, and during the normal g. The marks centered in an area about 2 in. to the upper left of the target. The task was repeated during a zero g trajectory and the marks centered in the same area.

Locomotion of Individual Humans Inside Space Vehicles

When space vehicles become large enough that men will have to move around inside them during normal operation, the problem of locomotion arises. Normal walking is impossible, of course, because without g, the friction we use in ordinary walking is not present.

The methods of locomotion for individuals in space vehicles that have been investigated during zero g are pushing off walls and floating across open areas, and walking with magnetic or suction cup shoes.

Free-Floating During Zero G

For travel inside space vehicles, simply pushing off walls, floors or ceiling and floating to the desired location, seems to be the most feasible and easiest means of locomotion. It takes very little time to learn to push with a low enough force that the hands and arms can stop the body without difficulty. The problem of getting one's center of gravity and line of thrust aligned and pointed towards the desired destination takes several trials to learn, but cannot be considered particularly difficult. One may start to tumble (Figure 1) caused by having the line of thrust pass through the body at a point lower than the center of gravity, but usually the rotation is slow enough to be easily controlled and stopped when a wall or other

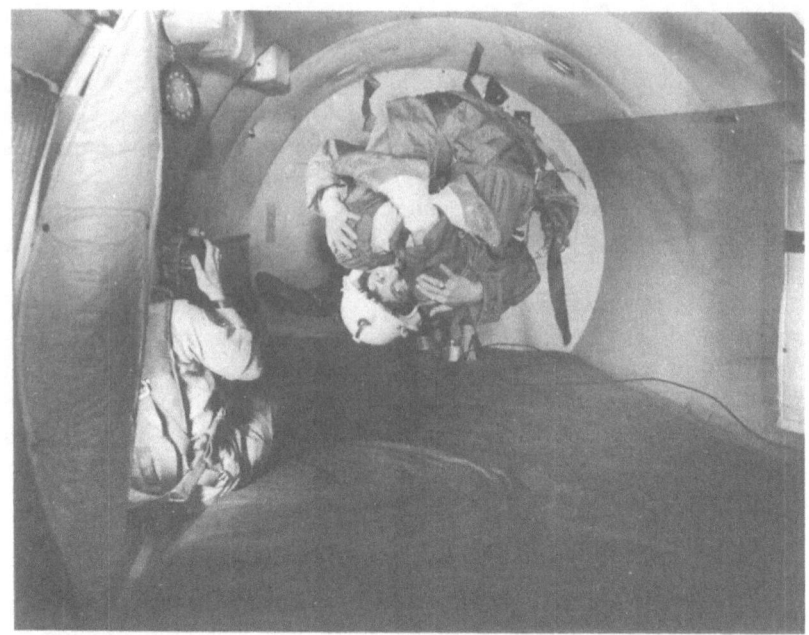

Figure 1. Experimentation on tumbling under zero gravity conditions.

solid object comes within reach. It appears that recessed handholds would be of considerable aid in this type of locomotion. As a safety measure for those employing this method of locomotion, it is recommended that sharp corners and other protuberances in the areas in which men are likely to travel by floating be padded. Of the methods tested so far, this free-floating appears to be the easiest and most carefree method of travel, not requiring mechanical aids, that has been devised by man.

The sensations of this free-floating are quite interesting. To the surprise of all who experience the complete zero g of free-floating, the sensation one gets when a fast elevator starts down is now present. This may be because the time we take to get from 1 g to zero g is considerably longer than the time a fast elevator takes. One has to keep pushing against the pad to test whether or not he is at zero. When a slight push starts one towards the ceiling, he knows he is at zero. Nearly everyone experiences a sense of exhilaration during zero g. It is a very enjoyable experience, and very relaxing. For the novice flyers, these enjoyable sensations are sometimes interrupted by extreme nausea, but because of the 2½ g's immediately preceding and following the zero g period, it is impossible to say that zero g will cause nausea. It is far more likely that it is the changing g, not the zero g, that causes the discomfort.

Walking During Zero G

The use of both the suction cup and magnetic shoes was investigated about the same time. During the first trials, the shoes with the suction cups employed one suction cup located on the heel of each shoe. While it was possible to walk on the ceiling during zero g using these shoes, it was very apparent that smaller suction cups on the forward portion of the sole were necessary. Without suction cups on the forward part of the shoes, when one applied force to his ankle to force the body forward in preparation for a step, the front part of the shoe pulled away from the surface and there was no force available to get the body started forward. Even with forward placed suction cups, walking was difficult. This is not considered to be a satisfactory method of providing positive contact with a surface to facilitate walking during zero g.

The magnetic shoes, or more properly, magnetic slippers, had permanent magnets attached to the soles (Figure 2). The first few times the shoes were used, the magnets in each shoe had about 5 lb of attraction. While it was possible to walk on the ceiling with these shoes during zero g, it was difficult. The attractive force was so light that it was very easy to push one's self away from the ceiling and lose contact with it completely. For our next model, magnets were installed that give an attraction of about 18 lb. This is expected to reduce the problem of pushing away from the ceiling. These have not yet been tried during zero g.

To decrease the amount of force required to separate the magnetic shoes from the ceiling, yet have enough force to hold one securely in position, we designed and built a pair of electromagnetic shoes. These shoes are powered by two flashlight batteries and have a microswitch located so that as soon as one lifts the heel, which is easy to do, the circuit is broken and the shoes will readily come off the surface. They have a potentiometer in the circuit so that the attraction can be varied from zero to 60 lb. These have just recently been completed and have not been tried during zero g conditions.

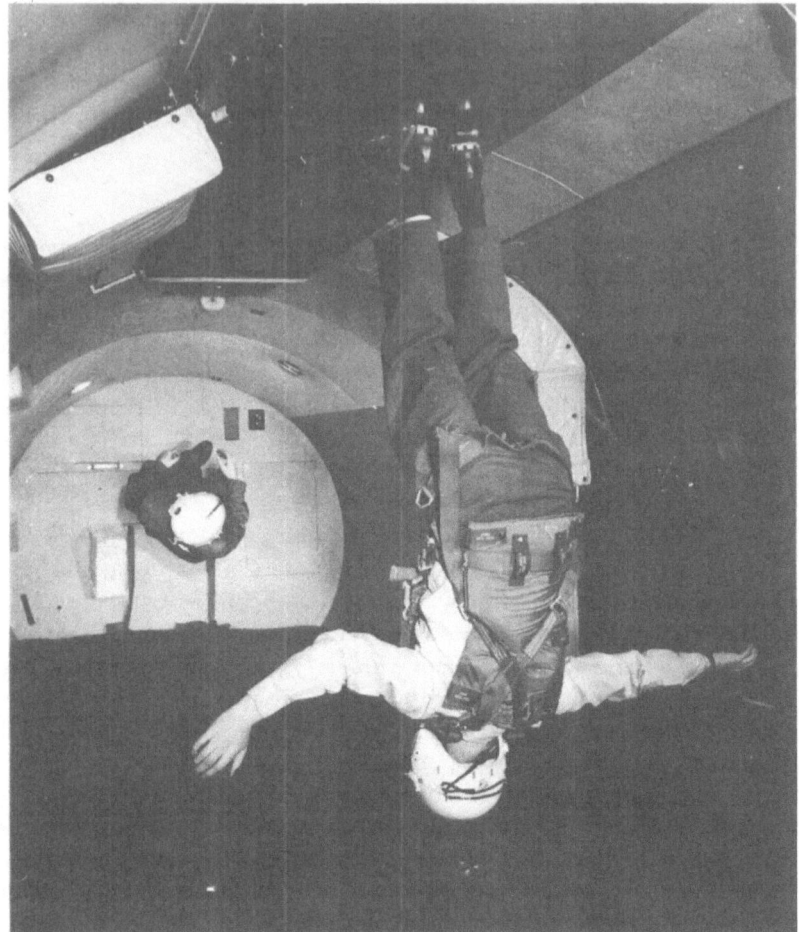

Figure 2. Use of magnetized soles for working in zero gravity environment.

Locomotion of Individual Humans Outside Space Vehicles

Locomotion of individuals outside their space vehicle will be required for tasks such as inspecting and repairing the exterior of their own vehicle, transferring supplies or personnel to adjacent orbiting vehicles, servicing unmanned communications and other satellites and for building and repairing large space stations.

For walking on the immediate surface of the vehicle, if the vehicle surface is ferrous or if magnetic particles are imbedded in pathways, the magnetic—or more preferably electromagnetic—shoes appear to be suitable. One would always have a rope attached to himself and the ship, of course, as a safety measure to prevent accidentally drifting off into space.

Locomotion of individuals outside space vehicles will probably be based on the use of a reaction propulsion type unit (Figure 3).

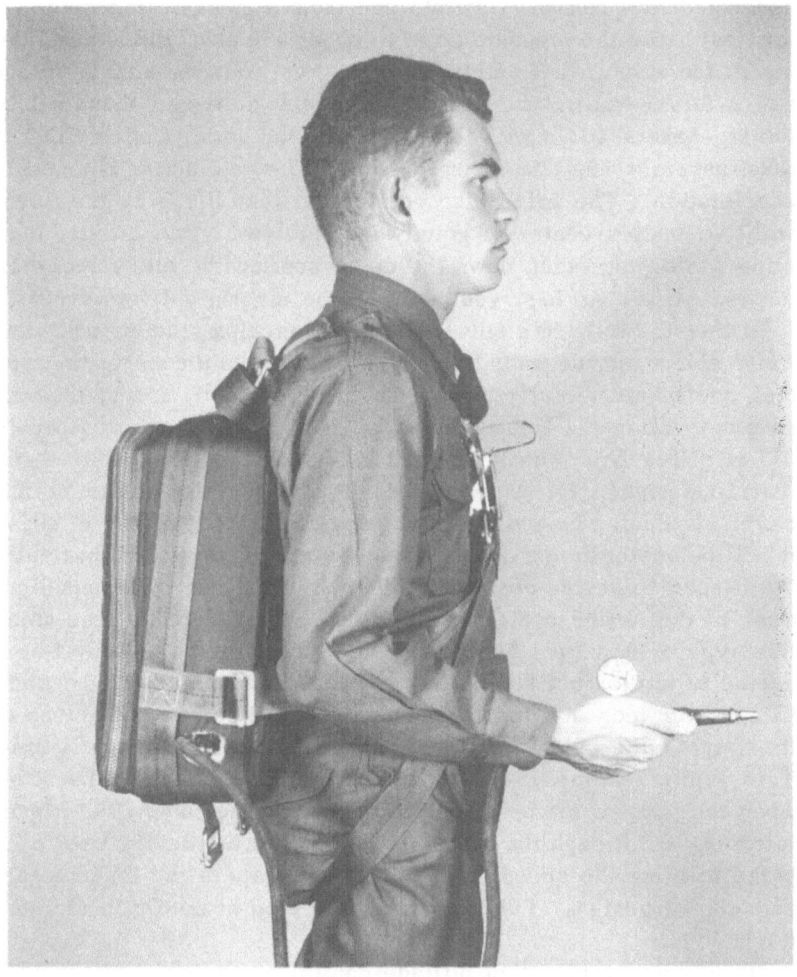

Figure 3. Pneumatic propulsion unit for locomotion in zero gravity environment.

This type of device can have a relatively low thrust and still be quite suitable because of the unique space flight conditions of no air and zero gravity.

The first reaction propulsion unit which was developed consisted of taking standard items of equipment that were immediately available and combining them to produce our unit. The source of energy was six steel bottles of air compressed to 1,800 psi. These are connected to a common manifold and then a pressure reducing system that takes the pressure down to about 450 psi. From there, the air is led through a flexible hose to a very simple nozzle with a button for controlling the flow. This system delivered about 5 lb of thrust. Several trials with this device during zero g indicate that it takes several times this amount of thrust to give a human reasonable acceleration. The trials also pointed out that lining up the thrust with the body's center of gravity is a problem. Uncontrolled tumbling while proceeding down the cabin seemed the rule rather than the exception. An improved model of the air gun delivers about 17 lb of thrust. While this solved the power problem, the attitude control problem was accentuated. Both of these units are of the type that one points directly away from the desired track and then attempts to align the line of thrust so it will go through the body's center of gravity. The latest unit we have ready for test is a gun that is designed for expelling air from either the breech or the barrel, depending on which of two triggers is actuated. It is hoped it will be easier to use a gun one points at the target and then pulls the trigger, releasing air from the breech of the gun. The misalignment of thrust and center of gravity is expected to be less using this type of unit than using a pusher type unit. The body is expected to stretch out behind the gun and small bursts can be made that may correct any tumbling tendencies. When one gets close to the target, one pulls the trigger releasing air through the barrel. This, while slowing one down, will also probably rotate the body. When the body is rotated about 180 deg from the target, the trigger releasing air through the breech is pulled and the device used as a brake to bring the speed down to the point where the legs can absorb the momentum. This unit will be tested at zero g in the near future.

It was anticipated that attitude control by the sole use of a reaction gun or body-mounted reaction units would be difficult. Accordingly experiments on the use of gyros to aid in body stabilization during zero g were performed (Figure 4).   The unit tested for its effect on stabilizing a human body during zero g consisted of a 16-lb, 9-in. × 4-in. aluminum wheel mounted on a 15-in. axle and enclosed in an aluminum box.   Just before the experimental runs at zero g, the unit was accelerated to about 3650 rpm by holding the aluminum gyro wheel against a rotating drum driven by an electric motor.   The experiments indicated that this gyro will stabilize a human subject during zero g.   Subjects have been stabilized by the gyro during both prone and squatting positions.

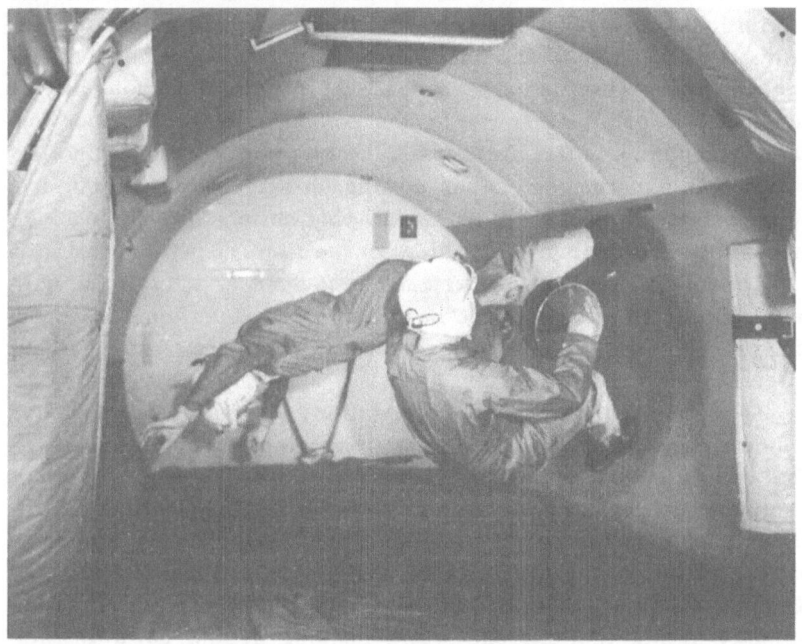

Figure 4. Experiment on gyroscopic stabilization of the human body under zero gravity conditions.

Future work in this area will be aimed at decreasing the size of the gyro unit, possibly using two mounted at right angles to one another, and developing motors and sensing units to drive the gyros about their various axis in order to stabilize the human's attitude while being propelled by rocket propulsion units or when attitude stabilization is desired.

Human Perceptive Orientation During Zero G

When zero g trajectories were first initiated (see Figure 4), it was mildly surprising to find that visual orientation was no problem. "Up" stayed up and "down" stayed down, with no tendency for the two to become confused. Even when free-floating locomotion from one point in the cabin to another began during zero g, it was apparent that visual orientation was no problem. To explore this problem further the author performed some tumbling maneuvers in which, by appropriate movements of the arms and legs, he was able to set his body in rotation, head over heels, with a speed of about 1 to 1.5 revolutions per sec. While it was difficult to achieve this rotation within the confines of the cabin without contacting either the ceiling or floor, after practice it became possible occasionally to achieve five or six turns during the zero g trajectory without contacting anything. After about two of these turns, complete disorientation was experienced. When coming up against a solid surface, it was extremely difficult to determine what surface one was against. It could be the ceiling, either side, or the floor. The disorientation was complete. Orientation was not recovered until after the pull-out started, and the subject had returned to the floor and spent several seconds there, being crushed into the mattress by the g forces of the pull-out.

The next step in this study of orientation came quite unexpectedly; it occurred when ceiling walking was initiated. When one walks on the ceiling, using either the suction cups or magnetic shoes, an overpowering sensation of the ceiling becoming the floor is created. Other people in the cabin oriented to the conventional "up-down" appear to the ceiling walker to be upside down. When one looks to the front of the airplane, one is quite surprised to see the pilots apparently flying the airplane while hanging upside down in their seats. From this evidence, it is judged that perceptual orientation in an orbiting or space vehicle may not be a difficult problem. It is quite probable that a tight safety belt, holding one's buttocks firmly against a seat, coupled with the normal visual up-down associated with one's instrument panel, will be sufficient to provide an acceptable and comfortable up-down relationship for pilots.

Behavior of Liquids During Zero G

Studying the behavior of liquids during the 15 sec zero g periods is quite a problem. If the plane is brought slowly to zero from a positive g condition, nothing will change unless a mechanical force of some kind is imparted to the liquid. However, bring the plane to an absolute zero g both vertically and horizontally is extremely difficult, and occurs quite infrequently. One method of achieving a near absolute zero g made use of a sphere, containing the experiment, which was suspended by cables and chains from a bridgelike structure that could be motor-driven up and down the cabin to neutralize longitudinal accelerations of the airplane. The plan was to slack off on the cables and chains when the plane first achieved zero, and then by driving the retaining structure up and down the cabin, allow the sphere to stay at zero but still be restrained by the cables so as to maintain control and keep it from crashing into the sides or top of the cabin.

In practice, however, this system did not work. It was found that the momentum caused by the rapidly slackening chains was sufficient to impart an acceleration to the sphere, thus voiding the reason for having the free-floating sphere within the retaining framework. The best results are obtained when the sphere is turned loose in the aft portion of the cabin, and several men, experienced in handling themselves at zero g, monitor the behavior of the free-floating sphere and see that it does not damage itself or the airplane. This method has been adopted for use in the KC-135 and appears to be the most feasible method for obtaining zero g for periods of more than two or three seconds. Until we have a research laboratory in orbit, this appears to be the best situation for studying the effects of zero g on liquids that we can achieve.

Fluid-Transfer Problems During Zero G

Fluid-transfer problems occur when starting liquid rocket engines during orbital flight or during the operation of auxiliary power systems during unpowered phases of orbital flight. The application of air pressure to a container and taking the fluid off from the bottom will not work. A positive mechanical pressure must be applied to the fluid. This problem was dramatically brought to attention during the early flights by the propellers overspeeding. The propellers on the C-131B are controlled through the use of oil pressure. The oil is picked up in the bottom of the engine crankcase by a sump pump and piped to the propeller. During one of the early flights, the plane was put into slightly negative g, causing all the oil to leave the bottom of the crankcase section. The plane did not return to positive g, so the oil did not return to the sump pump, very shortly causing a drop in oil pressure at the propeller, which resulted in overspeeding. The solution of this problem involved the installation of a rather complicated system that used nitrogen pressure operating against a flexible diaphram that forced oil into the lines during the zero g portion of the maneuver. While the system is a bit cumbersome to use, it does solve the problem of overspeeding propellers.

Other methods of transferring fluid at zero g involve the use of a cylinder-type container and a device for causing the fluid mass to rotate rapidly in such a manner that it forces itself against the inner circumference of the cylinder. The fluid can then be picked off from any point on the circumference and air bled into the center of the tank.

Heat Transfer Problems During Zero G

Those who are familiar with standard heat-transfer equations for liquid systems will recall that g is involved in the equation. Not only is it in the equation, but it is involved in the action to the extent that when it drops to zero, the heat transferred drops to zero. Obviously, when dealing with the problem of heat transfer during zero g, the basic equations must be modified or means found to introduce an artificial g to the heat transfer system. One apparatus for experiments designed to help define the problem area was found to be unsuitable on one of the first flights, so it was to be redesigned. For the latest information on this area, contact the Accessories Laboratory, Wright Air Development Division.

Conclusions

As a result of the experience gained by the Crew Stations Section while flying over 2000 zero g trajectories, the following information about human behavior and performance during zero g is believed to be correct. Floating about in the padded cabin during a zero g maneuver is a very exhilarating experience. Free-floating, without restraints of any kind, is quite different from going through the maneuver while strapped down. The feeling has been reported by various people as being "floating," "dry swimming" and just plain "fun." Learning to maneuver the body and proceed to predetermined points takes a little practice, but can be managed after a few trials. Getting the center of gravity lined up with the line of thrust, and the line of thrust aimed at the intended point are the major problems. Judging the amount of push to give to proceed to the desired point does not seem to be very difficult.

While rigorously scientific tests of motor performance during zero g have not been completed, so far no significant decrement in performance due to zero g appears likely, provided the subject is properly strapped down so as to have a fixed position to work from, without sliding about as a result of body movements. Speed and accuracy of hand motions appear to be about the same during zero g as during normal g. In fact, moving the arm through a large arc is reported to be easier during zero g than during normal g.

Several observations have been made concerning the problems of orientation during zero g. It appears that when strapped in a seat or floating about in the cabin without tumbling, "up" stays up and "down" stays down. When tumbling, there seems to be no "up" or "down." Tumbling seems to produce complete disorientation, so complete that it takes several seconds of positive g at the completion of the maneuver for one to become reoriented again. The feeling of disorientation is so marked that it could be said to border on a severe case of vertigo. When walking on the ceiling, the curious phenomena of the ceiling turning into the floor has been reported by all five people who have performed this maneuver. This inversion takes place about the time the feet first become attached to the ceiling, either with the suction cup or the magnetic shoes, and reverses itself when the maneuver is completed and the subject is pulled off the ceiling and returns to the normal padded floor. The subject, when walking on the ceiling, has the distinct and powerful feeling that everyone else in the airplane is upside down, "even those pilots in the cockpit who are flying the airplane while hanging upside down in their seats."

The studies conducted thus far on individual propulsion indicate that the human engineering problems of designing a propulsion system that a man can use to maneuver about while assembling a space station, or while transferring from one vehicle to another while in space, are going to be of considerable size. The tested system, while it provides a thrust that is considered marginal, demonstrates that uncontrolled tumbling occurs when a pusher type, hand-held unit is employed as the source of thrust. This study will be continued until the human engineering requirements for a rocket-type propulsion unit, together with the appropriate controls and displays for controlling individual propulsion in a space environment, can be specified.

The problems of handling liquid during zero g are difficult to work on due to the extreme shortness of available zero g periods, but appear to be amenable to straightforward engineering approaches. The problems of heat transfer using liquid systems appear a little more difficult, but the Accessories Laboratory at the Wright Air Development Division had several ideas for dealing with them and is in the process of defining the problem and working on it at this time.